R-101-2002

Recommended Practice

The CFD General Notation System – Standard Interface Data Structures

Maintained by
The CGNS Steering Sub-committee of the AIAA CFD Committee on Standards

Abstract

The CFD General Notation System (CGNS) is a standard for recording and recovering computer data associated with the numerical solution of the equations of fluid dynamics. The intent is to facilitate the exchange of CFD data between sites, between applications codes, and across computing platforms, and to stabilize the archiving of CFD data.

The CGNS system consists of a collection of conventions, and software implementing those conventions, for the storage and retrieval of CFD data. It consists of two parts: (1) a standard format for recording the data, and (2) software that reads, writes, and modifies data in that format. The format is a conceptual entity established by the documentation; the software is a physical product supplied to enable developers to access and produce data recorded in that format.

The standard format, or paper convention, part of CGNS consists of two fundamental pieces. The first, known as the Standard Interface Data Structures, is described in this Recommended Practice. It defines the intellectual content of the information to be stored. The second, known as the File Mapping, defines the exact location in a CGNS file where the data is to be stored.

Recommended practice: the CFD general notation system – standard interface data structures / maintained by the CGNS Steering Sub-committee of the AIAA CFD Committee on Standards.

   p. cm.

"R-101-2002."

ISBN 1-56347-558-8 (softcover) – ISBN 1-56347-559-6 (electronic)

   1. Fluid dynamics – Data processing – Standards. 2. Fluid dynamics -- Mathematics. 3. Numerical analysis. I. American Institute of Aeronautics and Astronautics.

QA911 .R42 2002
620.1'064'0285 – dc21

2002034286

Published by

American Institute of Aeronautics and Astronautics
1801 Alexander Bell Drive, Reston, VA 20191

Printed in the United States of America.

Contents

List of Figures

List of Tables

Foreward

This document contains the Standard Interface Data Structures (SIDS) definitions for the CFD General Notation System (CGNS) project. This project was originally a NASA-funded contract under the AST program, but control has now been completely transferred to a public forum known as the CGNS Steering Committee, a sub-committee of the AIAA CFD Committee on Standards.

This is Version 2.0 beta 2, Revision 1, of this document, released 9 Feb 2001. The current version was derived from the 11 Aug 1999 draft written by Steve Allmaras at Boeing.

The purpose of this document is to scope the information that should be communicated between various CFD application codes; the target is 3–D multizone, compressible Navier-Stokes analysis. Attention in this document is not focussed on I/O routines or formats, but on the precise description of data that should be present in the I/O of a CFD code or in a CFD database.

This document therefore contains a precise definition of information pertinent to a CGNS database. Specifically, the following information is addressed:

- grid coordinates and elements

- flow solution data, including nondimensional parameters

- multizone interface connectivity, including abutting and overset

- boundary conditions

- flow equation descriptions

- time-dependent flow

- reference states

- dimensional units and nondimensionalization information associated with data

- convergence history

- association to geometry definition

- topologically based hierarchical structures

This information is encoded into C-like data structures.

The AIAA Standards Program Procedures provide that all approved Standards, Recommended Practices, and Guides are advisory only. Their use by anyone is entirely voluntary. There is no advance agreement to adhere to any AIAA standard publication and no commitment to conform to or be guided by any standards report. In formulating, revising, and approving standards publications, the Committees on Standards are responsible for protecting themselves against liability for infringement of patents or copyrights or both. This AIAA document is a voluntary standard. It becomes binding only when two parties agree to use it or parts thereof in a contract.

The CGNS Steering Committee, serving as the consensus body for this project, approved the document in March 2002. At the time of approval, the committee was composed of the following individuals:

Bob Bush, Chairman	Pratt & Whitney
Theresa Benyo	NASA Glenn
John Chawner	Pointwise, Inc.
Alexandre Corjon	Aerospatiale Matra Airbus
Ray Cosner	Boeing Phantom Works
Armen Darian	Boeing Space & Communications
Francis Enomoto	NASA Ames
Steve Feldman	CD Adapco Group
Steve Legensky	Intelligent Light
Doug McCarthy	Boeing Commercial
Marc Poinot	ONERA
Diane Poirier	ICEM CFD Engineering
Greg Power	Sverdrup, AEDC
Chris Rumsey	NASA Langley
Dave Schowalter	Fluent, Inc.
Marc Tombroff	NUMECA International
Kurt Weber	Rolls-Royce Allison

The AIAA Standards Executive Council accepted this document for publication in April 2002.

Questions and comments on this document are welcome and should be directed to one of the following:

Charlie Towne
MS 86-7
NASA Glenn Research Center
Cleveland, OH 44135-3191
(216) 433-5851
(216) 977-7500 (FAX)
e-mail: towne@grc.nasa.gov

Diane Poirier
ICEM CFD Engineering
2855 Telegraph Ave Suite 501
Berkeley, CA 94705
(510) 549-1890
(510) 841-8523 (FAX)
e-mail: diane@icemcfd.com

1 Introduction

The CGNS (CFD General Notation System) project originated during 1994 through a series of meetings that addressed improved transfer of NASA technology to industry. A principal impediment in this process was the disparity in I/O formats employed by various flow codes, grid generators, and other utilities, and CGNS was conceived as a means to promote "plug-and-play" CFD. Agreement was reached to develop CGNS at Boeing, under NASA Contract NAS1-20267, with active participation by a team of CFD researchers from NASA's Langley, Lewis (now Glenn), and Ames Research Centers, McDonnell Douglas Corporation (now part of Boeing), and Boeing Commercial Airplane Group. This team, which was joined by ICEM CFD Engineering Corporation of Berkeley, California in 1997, undertook the core of the development. However, in the spirit of creating a completely open and broadly accepted standard, all interested parties were encouraged to participate; the US Air Force and Arnold Engineering Development Center were notably present. From the beginning, the purpose was to develop a system that could be distributed freely, including all documentation, software and source code. This goal has now been fully realized; further, control of CGNS has been completely transferred to a public forum known as the CGNS Steering Committee.

The specific purpose of CGNS was to provide a standard for recording and recovering computer data associated with the numerical solution of the equations of fluid dynamics. The intent was to facilitate the exchange of CFD data between sites, between applications codes, and across computing platforms, and to stabilize the archiving of CFD data. The format implemented by this standard was to be (1) general, (2) portable, (3) expandable, and (4) durable.

The resulting system today consists of a collection of conventions, and software implementing those conventions, for the storage and retrieval of CFD data. The system consists of two parts: (1) a standard format for recording the data, and (2) software that reads, writes, and modifies data in that format. The format is a conceptual entity established by the documentation; the software is a physical product supplied to enable developers to access and produce data recorded in that format.

The principal target is the data normally associated with compressible viscous flow (i.e., the Navier-Stokes equations), but the standard is also applicable to subclasses such as Euler and potential flows. The initial release addressed multi-zone grids, flow fields, boundary conditions, and zone-to-zone connection information, as well as a number of auxiliary items, such as non-dimensionalization, reference states, and equation set specifications. Extensions incorporated since then include unstructured mesh, connections to geometry definition, and time-dependent flow.

It is worth noting that extensibility is a fundamental design characteristic of the system, which in principal could be used for other disciplines of computational field physics, such as acoustics or electromagnetics, given the willingness of the cognizant scientific community to define the conventions.

The standard format, or paper convention, part of CGNS consists of two fundamental pieces. The first, known as the Standard Interface Data Structures (SIDS), describes in detail the intellectual content of the information to be stored. It defines, for example, the precise meaning of a "boundary condition". The second, known as the File Mapping, defines the exact location in a CGNS file where the data is to be stored.

The implementation, or software, part of CGNS likewise consists of two separate entities. CGNS files are read and written by a stand-alone database manager called ADF (Advanced Data Format).

ADF manages a tree-like data structure, implemented as a binary file. Since the format of this file is completely controlled by ADF, and since ADF is written in ANSI C (FORTRAN wrappers are provided), these files and ADF itself are portable to any environment which supports ANSI C. ADF is available separately and constitutes a useful tool for the storage of large quantities of scientific data.

ADF, however, implements no knowledge of CFD or of the File Mapping. To simplify access to CGNS files, a second layer of software known as the Mid-Level Library is provided. This layer is in effect an API, or Application Programming Interface for CFD. The API incorporates knowledge of the CFD data structures, their meaning and their location in the file, enabling applications such as flow codes and grid generators to access the data in familiar terms. The API is therefore the piece of the CGNS system most visible to applications developers. Like ADF, the API is written in ANSI C; all public API routines have FORTRAN counterparts.

This document presents the formal definition of the Standard Interface Data Structures (SIDS). Section 2 presents the major design philosophies used to develop the CGNS database and the encoding of this database into the SIDS; this section also provides an overview of the database hierarchy. Section 3 describes the C-like nomenclature conventions used to define the SIDS. This section also gives the conventions for structured grid indexing and unstructured element numbering, and the nomenclature for multizone interfaces. Low-level building-block structures are defined in Section 4; these structures are used to define all higher-level structures. Structures for defining data arrays, including dimensional-units and nondimensional information, are presented in Section 5. The top levels of the CGNS hierarchy are next defined in Section 6. The following sections then fill out the remainder of the hierarchy: Section 7 defines the grid-coordinate, elements, and flow-solution structures; Section 8 defines the multizone interface connectivity structures; Section 9 defines boundary-condition structures; Section 10 defines structures for describing governing flow equations; Section 11 defines structures related to time-dependent flows; and Section 12 contains miscellaneous structures. Two appendices complete the document. Annex A provides naming conventions for data contained within the CGNS database, and Annex B contains a complete SIDS description of a structured-grid two-zone test case.

1.1 Major Differences from Previous SIDS Documents

The following items represent noteworthy alterations and additions to the SIDS starting with the August 1999 draft document. (Note that some of these changes — notably those for unstructured zones, family, and geometry reference — have existed previously in separate documents, but are now being merged officially into the SIDS; the data structures themselves are not "new.")

1.1.1 Version 2.0 (Beta 1)

The following changes were made for version 2.0 (beta 1) of the SIDS.

- The capability for recording unstructured zones has been added to the SIDS. (These changes occur throughout the document, although some specific items are listed below.)

- The values `UserDefined` and `Null` are now allowed for all enumeration types (throughout document).

- The following nodes are now defined (some of these also include additional new children subnodes): `Family_t` (Section 12.6), `Elements_t` (Section 7.3), `ZoneType_t` (Section 6.2), `FamilyName_t` (Section 6.2), `GeometryReference_t` (Section 12.7), `FamilyBC_t` (Section 12.8).

- Under `CGNSBase_t`, the `IndexDimension` is no longer recorded; it has been replaced by `CellDimension` and `PhysicalDimension` (Section 6.1).

- Under `Zone_t`, the optional parameter `VertexSizeBoundary` has been added for unstructured zones (Section 6.2).

- The method for general connectivity (`GridConnectivity_t`) has been altered. It now requires the use of either (a) `PointListDonor` (an integer, for `Abutting1to1` only) or (b) `CellListDonor` (an integer) plus `InterpolantsDonor` (a real) (Section 8.4).

- The `GridLocation_t` parameter has been moved up one level (from `BCDataSet_t` to `BC_t`). Thus, for example, if the boundary conditions are defined at vertices (the default), then any associated dataset information must also be specified at vertices (Section 9.3 and Section 9.4).

- The data-name identifier `LengthReference` has been added (Section 12.1 and Annex A.2).

- The ν_t parameter has been renamed `ViscosityEddyKinematic`, and a new parameter `ViscosityEddy`, representing μ_t, has been defined (Annex A.2).

1.1.2 Version 2.0 (Beta 2)

The following changes were made for version 2.0 (beta 2) of the SIDS.

- The following data structures related to time-dependent flow have been added: `BaseIterativeData_t` (Section 11.1.1), `ZoneIterativeData_t` (Section 11.1.2), `RigidGridMotion_t` (Section 11.2), `ArbitraryGridMotion_t` (Section 11.3).

2 Design Philosophy of Standard Interface Data Structures

The major design goal of the SIDS is a comprehensive and unambiguous description of the 'intellectual content' of information that must be passed from code to code in a multizone Navier-Stokes analysis system. This information includes grids, flow solutions, multizone interface connectivity, boundary conditions, reference states and dimensional units or normalization associated with data.

Implications of CFD Data Sets

The goal is description of the data sets typical of CFD analysis, which tend to contain a small number of extremely large data arrays. This has a number of implications for both the design of the SIDS and the ultimate physical files where the data resides. The first is that any I/O system built for CFD analysis must be designed to efficiently store and process large data arrays. This is reflected in the SIDS, which includes provisions for describing large data arrays.

The second implication is that the nature of the data sets allows for thorough description of the data with relatively little storage overhead and performance penalty. For example, the flow solution of a CFD analysis may contain several millions of quantities. Therefore, with little penalty it is possible to include information describing the flow variables stored, their location in the grid, and dimensional units or nondimensionalization associated with the data. The SIDS take advantage of this situation and includes an extensive description of the information stored in the database.

The third implication of CFD data sets is that files containing a CFD database are almost always required to be binary – ASCII storage of CFD data involves excessive storage and performance penalties. This means the files are not readable by humans and the information contained in them is not directly modifiable by text editors and such. This is reflected in the syntax of the SIDS, which tends to be verbose and thorough; whereas, directly modifiable ASCII file formats would tend to foster a more brief syntax.

It is important to note that the description of information by the SIDS is independent of physical file formats. However, it is targeted towards implementation using the ADF Core library. Some of the language components used to define the SIDS are meant to directly map into elements of an ADF node.

Topologically Based Hierarchical Database

An early decision in the CGNS project was that any new CFD I/O standard should include a hierarchical database, such as a tree or directed graph. The SIDS describe a hierarchical database, precisely defining both the data and their hierarchical relationships.

There are two major alternatives to organizing a CFD hierarchy: topologically based and data-type based. In a topologically based graph, overall organization is by zones; information pertaining to a particular zone, including its grid coordinates or flow solution, hangs off the zone. In a data-type based graph, organization is by related data. For example, there would be two nodes at the same level, one for grid coordinates and another for the flow solution. Hanging off each of these nodes would be separate lists of the zones.

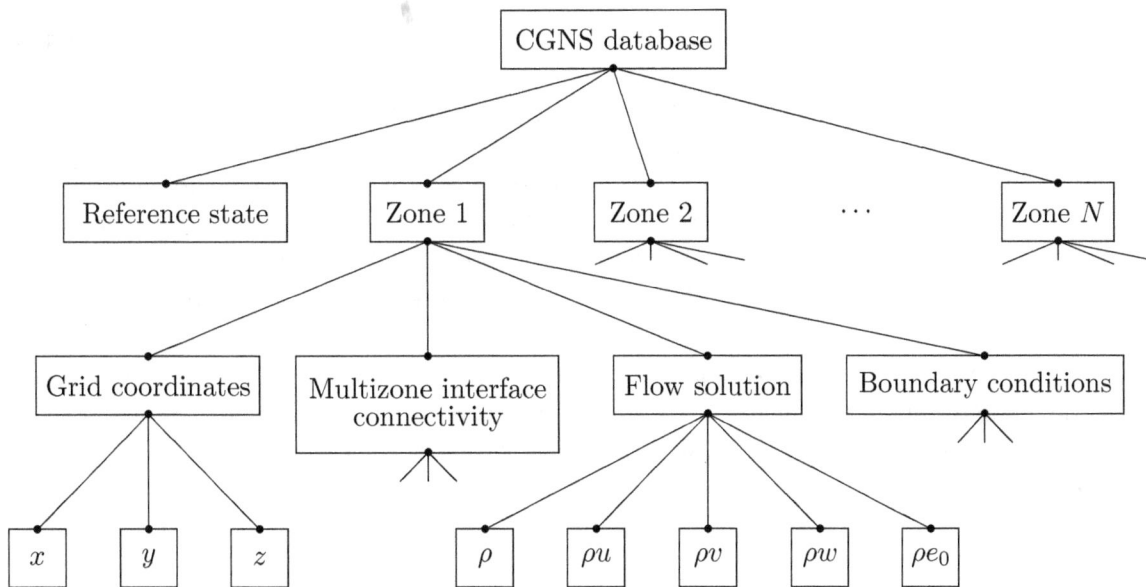

Figure 1: Sample Topologically Based CFD Hierarchy

The hierarchy described in this document is topologically based; a simplified illustration of the database hierarchy is shown in Figure 1. Hanging off the root 'node' of the database is a node containing global reference-state information, such as freestream conditions, and a list of nodes for each zone. The figure shows the nodes that hang off the first zone; similar nodes would hang off of each zone in the database. Nodes containing the physical-coordinate data arrays (x, y and z) for the first zone are shown hanging off the 'grid coordinates' node. Likewise, nodes containing the first zone's flow-solution data arrays hang off the 'flow solution' node. The figure also depicts nodes containing multizone interface connectivity and boundary condition information for the first zone; subnodes hanging off each of these are not pictured.

Additional Design Objectives

The data structures comprising the SIDS are the result of several additional design objectives:

One objective is to minimize duplication of data within the hierarchy. Many parameters, such as the grid size of a zone, are defined in only one location. This avoids implementation problems arising from data duplication within the physical file containing the database; these problems include simultaneous update of all copies and error checking when two copies of a data quantity are found to be different. One consequence of minimizing data duplication is that information at lower levels of the hierarchy may not be completely decipherable without access to information at higher levels. For example, the grid size is defined in the zone structure (see Section 6.2), but this parameter is needed in several substructures to define the size of grid and flow-solution data arrays. Therefore, these substructures are not autonomous and deciphering information within them requires access to information contained in the zone structure itself. The SIDS must reflect this cascade of information within the database.

Another objective is elimination of nonsensical descriptions of the data. The SIDS have been carefully developed to avoid data qualifiers and other optional descriptive information that could be inconsistent. This has led to the use of specialized structures for certain CFD-related information. One example is a single-purpose structure for defining physical grid coordinates of a zone. It is possible to define the grid coordinates, flow solution and any other field quantities within a zone by a generic discrete-data structure. However, this requires the generic structure to include information defining the grid location of the data (e.g. the data is located at vertices or cell centers). Using the generic structure to describe the grid coordinates leads to a possible inconsistency. By definition the physical coordinates that define the grid are located at vertices, and including an optional qualifier that states otherwise makes no sense.

A final objective is to allow documentation inclusion throughout the database. The SIDS contain a uniform documentation mechanism for all major structures in the hierarchy. However, this document establishes no conventions for using the documentation mechanism.

3 Conventions

3.1 Data Structure Notation Conventions

The intellectual content of the CGNS database is defined in terms of C-like notation including typedefs and structures. The database is made up of entities, and each entity has a type associated with it. Entities include such things as the dimensionality of the grid, an array of grid coordinates, or a zone which contains all the data associated with a given region. Entities are defined in terms of types, where a type can be an integer or a collection of elements (a structure) or a hierarchy of structures or other similar constructs.

The terminology 'instance of an entity' is used to refer to an entity that has been assigned a value or whose elements have been assigned values. The terminology 'specific instance of a structure' is also used in the following sections. It is short for an instance of an entity whose type is a structure.

Names of entities and types are constructed using conventions typical of *Mathematica*[1]. Names or identifiers contain no spaces and capitalization is used to distinguish individual words making up a name; names are case-sensitive. The characters '.' and '/' should be avoided in names as these have special meaning when referencing elements of a structure entity.

The following notational conventions are employed:

!	comment to end of line
_t	suffix used for naming a type
;	end of a definition, declaration, assignment or entity instance
=	assignment (takes on the value of)
:=	indicates a type definition (typedef)
[ ]	delimiters of an array
{ }	delimiters of a structure definition
{{ }}	delimiters of an instance of a structure entity
< >	delimiters of a structure parameter list
int	integer
real	floating-point number
char	character
bit	bit
Enumeration( )	indicates an enumeration type
Data( )	indicates an array of data, which may be multidimensional
List( )	indicates a list of entities
Identifier( )	indicates an entity identifier
LogicalLink( )	indicates a logical link
/	delimiter for element of a structure entity
../	delimiter for parent of a structure entity
(r)	designation for a required structure element
(o)	designation for an optional structure element
(o/d)	designation for an optional structure element with default if absent

[1] *Mathematica 3.0*, Wolfram Research, Inc., Champaign, IL (1996)

An enumeration type is a set of values identified by names; names of values within a given enumeration declaration must be unique. An example of an enumeration type is the following:

```
Enumeration( Dog, Cat, Bird, Frog )
```

This defines an enumeration type which contains four values.

`Data()` identifies an array of given dimensionality and size in each dimension, whose elements are all of a given data type. It is written as,

```
Data( DataType, Dimension, DimensionValues[] ) ;
```

`Dimension` is an integer, and `DimensionValues[]` is an array of integers of size `Dimension`. `Dimension` and `DimensionValues[]` specify the dimensionality of the array and its size in each dimension. `DataType` specifies the data type of the array's elements; it may consist of one of the following: `int`, `real`, `char` or `bit`. For multidimensional arrays, FORTRAN indexing conventions are used. `Data()` is formulated to map directly onto the data section of an ADF node.

A typedef establishes a new type and defines it in terms of previously defined types. Types are identified by the suffix '`_t`', and the symbol '`:=`' is used to establish a type definition (or typedef). For example, the above enumeration example can be used in a typedef:

```
Pet_t := Enumeration( Dog, Cat, Bird, Frog ) ;
```

This defines a new type `Pet_t`, which can then be used to declare a new entity, such as,

```
Pet_t MyFavoritePet ;
```

By the above typedef and declaration, `MyFavoritePet` is an entity of type `Pet_t` and can have the values `Dog`, `Cat`, `Bird` or `Frog`. A specific instance of `MyFavoritePet` is setting it equal to one of these values (e.g. `MyFavoritePet = Bird`).

A structure is a type that can contain any number of elements, including elements that are also structures. An example of a structure type definition is:

```
Sample_t :=
  {
  int Dimension ;                                            (r)

  real[4] Vector ;                                           (o)

  Pet_t ObnoxiousPet ;                                       (o)
  } ;
```

where `Sample_t` is the type of the structure. This structure contains three elements, `Dimension`, `Vector` and `ObnoxiousPet`, whose types are `int`, `real[4]` and `Pet_t`, respectively. The type `int` specifies an integer, and `real[4]` specifies an array of reals that is one-dimensional with a length of four. The '(r)' and '(o)' notation in the right margin is explained below. Given the definition of `Sample_t`, entities of this type can then be declared (e.g. `Sample_t Sample1;`). An example of an instance of a structure entity is given by,

```
Sample_t Sample1 =
  {{
  Dimension = 3 ;
  Vector = [1.0, 3.45, 2.1, 5.4] ;
  ObnoxiousPet = Dog ;
  }} ;
```

Note the different functions played by single braces '{' and double braces '{{'. The first is used to delimit the definition of a structure type; the second is used to delimit a specific instance of a structure entity.

Some structure type definitions contain arbitrarily long lists of other structures or types. These lists will be identified by the notation,

```
List( Sample_t Sample1 ... SampleN ) ;
```

where `Sample1 ... SampleN` is the list of structure names or identifiers, each of which has the type `Sample_t`. Within each list, the individual structure names are user-defined.

In the CGNS database it is sometimes necessary to reference the name or identifier of a structure entity. References to entities are denoted by `Identifier()`, whose single argument is a structure type. For example,

```
Identifier(Sample_t) SampleName ;
```

declares an entity, `SampleName`, whose value is the identifier of a structure entity of type `Sample_t`. Given this declaration, `SampleName` could be assigned the value `Sample1` (i.e. `SampleName = Sample1`).

It is sometimes convenient to directly identify an element of a specific structure entity. It is also convenient to indicate that two entities with different names are actually the same entity. We borrow UNIX conventions to indicate both these features, and make the analogy that a structure entity is a UNIX directory and its elements are UNIX files. An element of an entity is designated by '/'; an example is `Sample1/Vector`). The structure entity that a given element belongs to is designated '../' A UNIX-like logical link that specifies the sameness of two apparently different entities is identified by `LogicalLink()`; it has one argument. An example of a logical link is as follows: Suppose a specific instance of a structure entity contains two elements that are of type `Sample_t`; call them `SampleA` and `SampleB`. The statement that `SampleB` is actually the same entity as `SampleA` is,

```
SampleB = LogicalLink(../SampleA) ;
```

The argument of `LogicalLink()` is the UNIX-like 'path name' of the entity with which the link is made. In this document, `LogicalLink()` and the direct specification of a structure element via '/' and '../' are actually seldom used. These language elements are never used in the actual definition of a structure type.

Structure type definitions include three additional syntactic/semantic notions. These are parameterized structures, structure-related functions, and the identification of required and optional fields within a structure.

As previously stated, one of our design objectives is to minimize duplication of information within the CGNS database. To meet this objective, information is often stored in only one location of the hierarchy; however, that information is typically used in other parts of the hierarchy. A consequence of this is that it may not be possible to decipher all the information associated with a given entity in the hierarchy without knowledge of data contained in higher level entities. For example, the grid size of a zone is stored in one location (in Zone_t, see Section 6.2), but is needed in many substructures to define the size of grid and solution-data arrays.

This organization of information must be reflected in the language used to describe the database. First, parameterized structures are introduced to formalize the notion that information must be passed down the hierarchy. A given structure type is defined in terms of a list of parameters that precisely specify what information must be obtained from the structure's parent. These structure-defining parameters play a similar role to subroutine parameters in C or FORTRAN and are used to define fields within the structure; they are also passed onto substructures. Parameterized structures are also loosely tied to templates in C++.

Parameterized structures are identified by the delimiters < > enclosing the list of parameters. Each structure parameter in a structure-type definition consists of a type and an identifier. Examples of parameterized structure type definitions are:

```
NewSample_t< int Dimension, int Fred > :=
  {
  int[Dimension] Vector ;                                            (o)

  Pet_t ObnoxiousPet ;                                               (o)

  Stuff_t<Fred> Thingy ;                                             (o)
  } ;

Stuff_t< int George > :=
  {
  real[George] IrrelevantStuff ;                                     (r)
  } ;
```

NewSample_t and Stuff_t are parameterized structure types. Dimension and Fred are the structure parameters of NewSample_t. George is the structure parameter of Stuff_t. All structure parameters in this example are of type int. Thingy is a structure entity of type Stuff_t; it uses the parameter Fred to complete its declaration. Note the use of George and Fred in the above example. George is a parameter in the definition of Stuff_t; Fred is an argument in the declaration of an entity of type Stuff_t. This mimics the use of parameters in function definitions in C.

A second language feature required to cope with the cascade of information within the hierarchy is structure-related functions. For example, the size of an array within a given structure may be a function of one or more of the structure-defining parameters, or the array size may be a function of an optional field within the structure. No new syntax is provided to incorporate structure-related functions; they are instead described in terms of their return values, dependencies, and functionality.

An additional notation used in structure typedefs is that each element or field within a structure definition is identified as required, optional, or optional with a default if absent; these are designated by '(r)', '(o)', and '(o/d)', respectively, in the right margin of the structure definition. These designations are included to assist in implementation of the data structures into an actual database and can be used to guide mapping of data as well as error checking. 'Required' fields are those essential to the interpretation of the information contained within the data structure. 'Optional' fields are those that are not necessary but potentially useful, such as documentation. 'Defaulted-optional' fields are those that take on a known default if absent from the database.

In the example of `Sample_t` above, only the element `Dimension` is required. Both elements `Vector` and `ObnoxiousPet` are optional. This means that in any specific instance of the structure, only `Dimension` must be present. An alternative instance of the entity `Sample1` shown above is the following:

```
Sample_t Sample1 =
  {{
  Dimension = 4 ;
  }} ;
```

None of the entities and types defined in the above examples are actually used in the definition of the SIDS.

As a final note, the reader should be aware that the SIDS is a conceptual description of the form of the data. The actual locations of data in the file is determined in the ADF mapping. (See the CGNS *SIDS-to-ADF File Mapping Manual.*)

3.2 Structured Grid Notation and Indexing Conventions

A *grid* is defined by its vertices. In a 3-D structured grid, the volume is the ensemble of cells, where each cell is the hexahedron region defined by eight nearest neighbor vertices. Each cell is bounded by six faces, where each face is the quadrilateral made up of four vertices. An edge links two nearest-neighbor vertices; a face is bounded by four edges.

In a 2-D structured grid, the notation is more ambiguous. Typically, the quadrilateral area composed of four nearest-neighbor vertices is referred to as a cell. The sides of each cell, the line linking two vertices, is either a face or an edge. In a 1-D grid, the line connecting two vertices is a cell.

A *structured multizone grid* is composed of multiple regions called *zones*, where each zone includes all the vertices, cells, faces, and edges that constitute the grid in that region.

Indices describing a 3-D grid are ordered (i, j, k); (i, j) is used for 2-D and (i) for 1-D.

Cell centers, face centers, and edge centers are indexed by the minimum of the connecting vertices. For example, a 2-D cell center (or face center on a 3-D grid) would have the following convention:

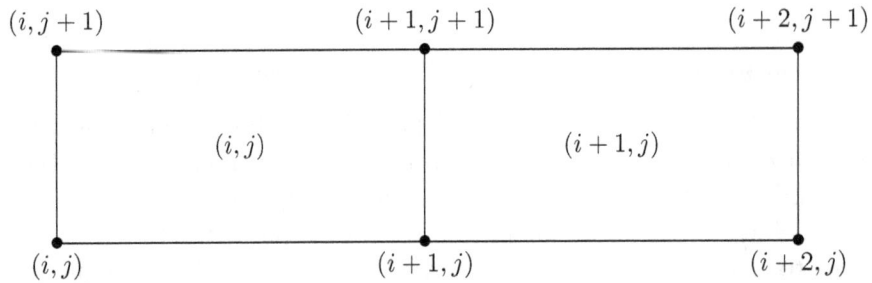

In addition, the default beginning vertex for the grid in a given zone is $(1, 1, 1)$; this means the default beginning cell center of the grid in that zone is also $(1, 1, 1)$.

A zone may contain grid-coordinate or flow-solution data defined at a set of points outside the zone itself. These are referred to as 'rind' or ghost points and may be associated with fictitious vertices or cell centers. They are distinguished from the vertices and cells making up the grid within the zone (including its boundary vertices), which are referred to as 'core' points. The following is a 2-D zone with a single row of 'rind' vertices at the minimum and maximum i-faces. The grid size (i.e. the number of 'core' vertices in each direction) is 5×4. 'Core' vertices are designated by '•', and 'rind' vertices by '×'. Default indexing is also shown for the vertices.

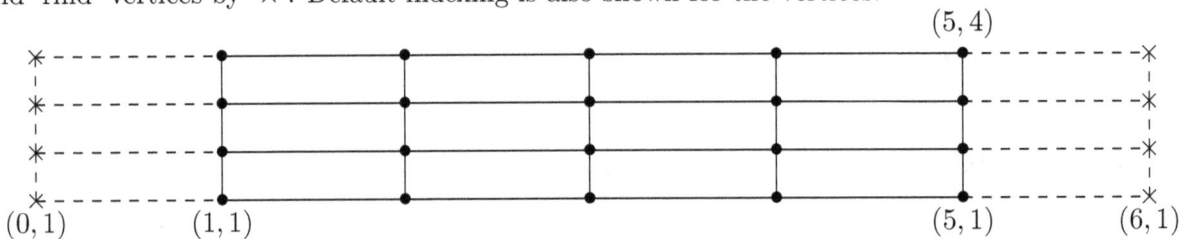

For a zone, the minimum faces in each coordinate direction are denoted i-min, j-min and k-min; the maximum faces are denoted i-max, j-max and k-max. These are the minimum and maximum 'core' faces. For example, i-min is the face or grid plane whose core vertices have minimum i index (which if using default indexing is 1).

3.3 Unstructured Grid Element Numbering Conventions

The major difference in the way structured and unstructured grids are recorded is the element definition. In a structured grid, the elements can always be recomputed easily using the computational coordinates, and therefore they are usually not written in the data file. For an unstructured grid, the element connectivity cannot be easily built, so this additional information is generally added to the data file. The element information typically includes the element type or shape, and the list of nodes for each element.

In an unstructured zone, the nodes are ordered from 1 to N, where N is the number of nodes in the zone. An element is defined as a group of one or more nodes, where each node is represented by its index. The elements are indexed from 1 to M within a zone, where M is the total number of elements defined for the zone.

CGNS supports eight element shapes — points, lines, triangles, quadrangles, tetrahedra, pentahedra, pyramids, and hexahedra. Elements describing a volume are referred to as 3-D elements.

Those defining a surface are 2-D elements. Line and point elements are called 1-D and 0-D elements, respectively.

In a 3-D unstructured mesh, the cells are defined using 3-D elements, while the boundary patches may be described using 2-D elements. The complete element definition may include more than just the cells.

Each element shape may have a different number of nodes, depending on whether linear or quadratic interpolation is used. Therefore the name of each type of element is composed of two parts; the first part identifies the element shape, and the second part the number of nodes. Table 1 summarizes the element types supported in CGNS.

Table 1: Element Types in CGNS

Dimensionality of the Element	Shape	Linear Interpolation	Quadratic Interpolation
0-D	Point	NODE	NODE
1-D	Line	BAR_2	BAR_3
2-D	Triangle	TRI_3	TRI_6
	Quadrangle	QUAD_4	QUAD_8, QUAD_9
3-D	Tetrahedron	TETRA_4	TETRA_10
	Pyramid	PYRA_5	PYRA_14
	Pentahedron	PENTA_6	PENTA_15, PENTA_18
	Hexahedron	HEXA_8	HEXA_20, HEXA_27

Any element type not supported by CGNS can be recorded using the CGNS generic element type NGON_n. See Section 7.3 for more detail.

The ordering of the nodes within an element is important. Since the nodes in each element type could be ordered in multiple ways, it is necessary to define numbering conventions. The following sections describe the element numbering conventions used in CGNS.

3.3.1 1-D (Line) Elements

1-D elements represent geometrically a line (or bar). The linear form, BAR_2, is composed of two nodes at each extremity of the line. The quadratic form, BAR_3, has an additional node located at the middle of the line.

BAR_2   BAR_3

1 ●————————————————● 2   1 ●————————●————————● 2
                                                              3

Face Definition

Oriented edge	Corner nodes	Mid-node
E1	N1,N2	N3

3.3.2   2-D (Surface) Elements

2-D elements represent a surface in either 2-D or 3-D space. Note that in physical space, the surface need not be planar, but may be curved. In a 2-D mesh the elements represent the cells themselves; in a 3-D mesh they represent faces. CGNS supports two shapes of 2-D elements — triangles and quadrangles.

The normal vector of a 2-D element is computed using the cross product of a vector from the first to second node, with a vector from the first to third node. The direction of the normal is such that the three vectors (i.e., $(\overrightarrow{N2} - \overrightarrow{N1})$, $(\overrightarrow{N3} - \overrightarrow{N1})$, and $\overrightarrow{N}$) form a right-handed triad.

$$\overrightarrow{N} = (\overrightarrow{N2} - \overrightarrow{N1}) \times (\overrightarrow{N3} - \overrightarrow{N1})$$

In a 2-D mesh, all elements must be oriented the same way; i.e., all normals must point toward the same side of the mesh.

3.3.2.1   Triangular Elements

Two types of triangular elements are supported in CGNS, TRI_3 and TRI_6. TRI_3 elements are composed of three nodes located at the three geometric corners of the triangle. TRI_6 elements have three additional nodes located at the middles of the three edges.

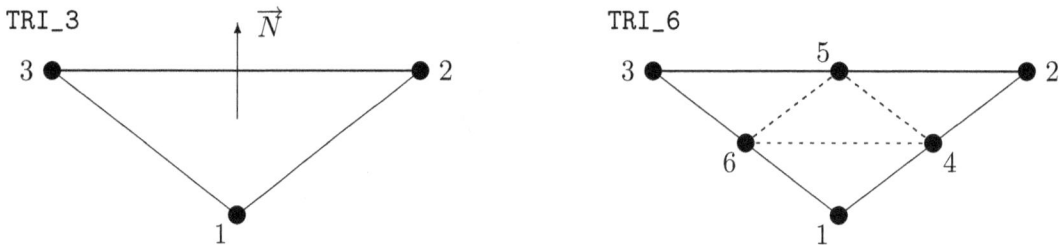

Edge Definition

Oriented edges	Corner nodes	Mid-node
E1	N1,N2	N4
E2	N2,N3	N5
E3	N3,N1	N6

Face Definition

Face	Corner nodes	Mid-edge nodes	Oriented edges
F1	N1,N2,N3	N4,N5,N6	E1,E2,E3

Notes

N1,...,N6	Grid point identification number. Integer ≥ 0 or blank, and N1 $\neq$ N2 $\neq \ldots \neq$ N6. Grid points N1, N2, and N3 are in consecutive order about the triangle.
E1,E2,E3	Edge identification number.
F1	Face identification number.

3.3.2.2 Quadrilateral Elements

CGNS supports three types of quadrilateral elements, QUAD_4, QUAD_8, and QUAD_9. QUAD_4 elements are composed of four nodes located at the four geometric corners of the quadrangle. In addition, QUAD_8 and QUAD_9 elements have four mid-edge nodes, and QUAD_9 adds a mid-face node.

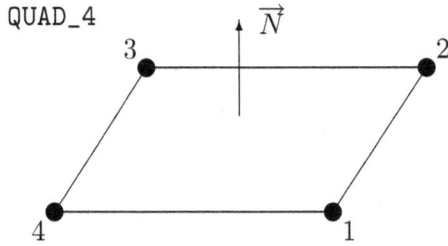

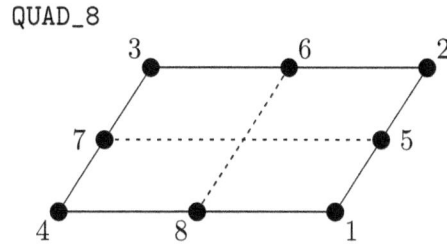

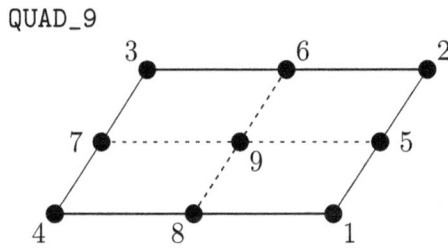

QUAD_4

QUAD_8

QUAD_9

Edge Definition

Oriented edges	Corner nodes	Mid-node
E1	N1,N2	N5
E2	N2,N3	N6
E3	N3,N4	N7
E4	N4,N1	N8

Face Definition

Face	Corner nodes	Mid-edge nodes	Mid-face node	Oriented edges
F1	N1,N2,N3,N4	N5,N6,N7,N8	N9	E1,E2,E3,E4

Notes

N1,...,N9   Grid point identification number. Integer ≥ 0 or blank, and N1 $\neq$ N2 $\neq \ldots \neq$ N9. Grid points N1...N4 are in consecutive order about the quadrangle.

E1,...,E4   Edge identification number.

F1   Face identification number.

3.3.3 3-D (Volume) Elements

3-D elements represent a volume in 3-D space, and constitute the cells of a 3-D mesh. CGNS supports four different shapes of 3-D elements — tetrahedra, pyramids, pentahedra, and hexahedra.

3.3.3.1 Tetrahedral Elements

CGNS supports two types of tetrahedral elements, TETRA_4 and TETRA_10. TETRA_4 elements are composed of four nodes located at the four geometric corners of the tetrahedron. TETRA_10 elements have six additional nodes, at the middle of each of the six edges.

TETRA_4

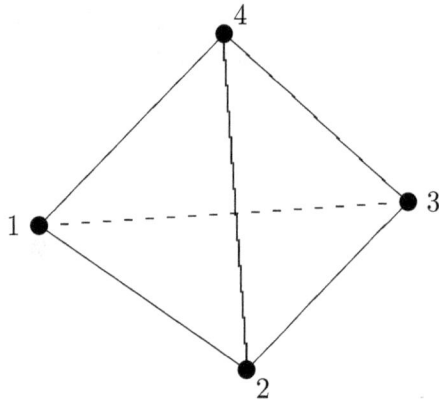

TETRA_10

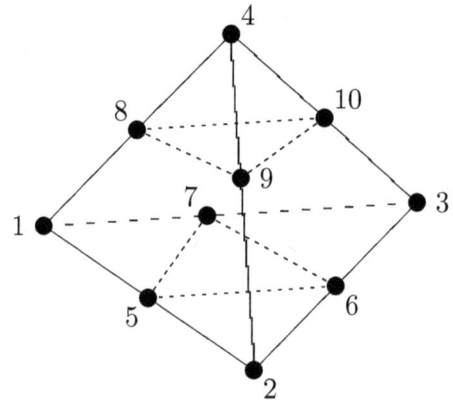

Edge Definition

Oriented edges	Corner nodes	Mid-node
E1	N1,N2	N5
E2	N2,N3	N6
E3	N3,N1	N7
E4	N1,N4	N8
E5	N2,N4	N9
E6	N3,N4	N10

Face Definition

Face	Corner nodes	Mid-edge nodes	Oriented edges
F1	N1,N3,N2	N7,N6, N5	-E3,-E2,-E1
F2	N1,N2,N4	N5,N9, N8	E1, E5,-E4
F3	N2,N3,N4	N6,N10,N9	E2, E6,-E5
F4	N3,N1,N4	N7,N8, N10	E3, E4,-E6

Notes

N1,...,N10   Grid point identification number. Integer ≥ 0 or blank, and N1 $\neq$ N2 $\neq$ ... $\neq$ N10. Grid points N1...N3 are in consecutive order about one trilateral face. The cross product of a vector going from N1 to N2, with a vector going from N1 to N3, must result in a vector oriented from face F1 toward N4.

E1,...,E6   Edge identification number. The edges are oriented from the first to the second node. A negative edge (e.g., -E1) means that the edge is used in its reverse direction.

F1,...,F4   Face identification number. The faces are oriented so that the cross product of a vector from its first to second node, with a vector from its first to third node, is oriented outward.

3.3.3.2 Pyramid Elements

CGNS supports two types of pyramid elements, PYRA_5 and PYRA_14. PYRA_5 elements are composed of five nodes located at the five geometric corners of the pyramid. PYRA_14 elements have nine additional nodes, eight located at the middle of each of the eight edges, and one at the cell center.

PYRA_5

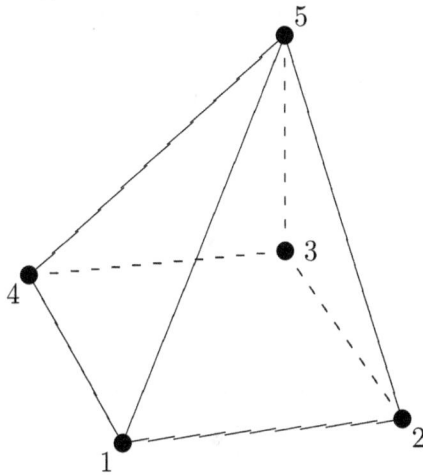

PYRA_14

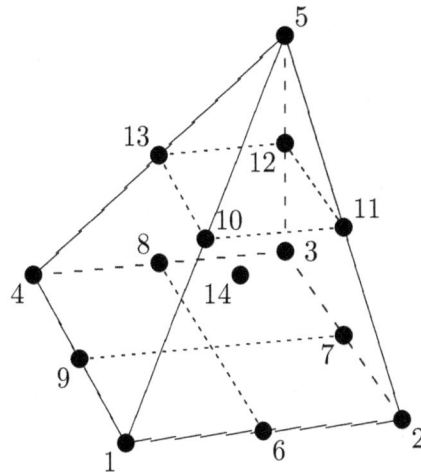

Edge Definition

Oriented edges	Corner nodes	Mid-node
E1	N1,N2	N6
E2	N2,N3	N7
E3	N3,N4	N8
E4	N4,N1	N9
E5	N1,N5	N10
E6	N2,N5	N11
E7	N3,N5	N12
E8	N4,N5	N13

Face Definition

Face	Corner nodes	Mid-edge nodes	Oriented edges
F1	N1,N4,N3,N2	N9,N8, N7, N6	-E4,-E3,-E2,-E1
F2	N1,N2,N5	N6,N11,N10	E1, E6,-E5
F3	N2,N3,N5	N7,N12,N11	E2, E7,-E6
F4	N3,N4,N5	N8,N13,N12	E6, E8,-E7
F5	N4,N1,N5	N9,N10,N13	E4, E5,-E8

Notes

N1,...,N14   Grid point identification number. Integer ≥ 0 or blank, and N1 $\neq$ N2 $\neq$ ...$\neq$ N14. Grid points N1...N4 are in consecutive order about the quadrilateral face. The cross product of a vector going from N1 to N2, with a vector going from N1 to N3, must result in a vector oriented from face F1 toward N5. N14 is located at the cell center.

E1,...,E8   Edge identification number. The edges are oriented from the first to the second node. A negative edge (e.g., -E1) means that the edge is used in its reverse direction.

F1,...,F5   Face identification number. The faces are oriented so that the cross product of a vector from its first to second node, with a vector from its first to third node, is oriented outward.

3.3.3.3 Pentahedral Elements

CGNS supports three types of pentahedral elements, PENTA_6, PENTA_15, and PENTA_18. PENTA_6 elements are composed of six nodes located at the six geometric corners of the pentahedron. In addition, PENTA_15 and PENTA_18 elements have a node at the middle of each of the nine edges; PENTA_18 adds a node at the middle of each of the three quadrilateral faces.

PENTA_6

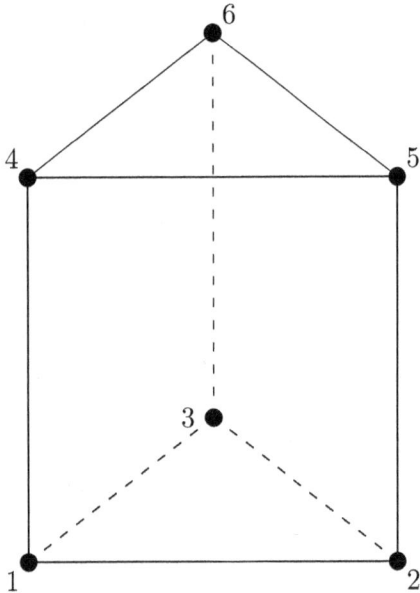

PENTA_15

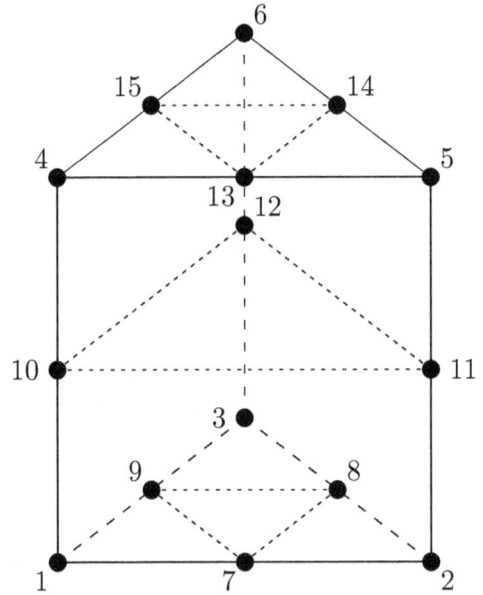

PENTA_18

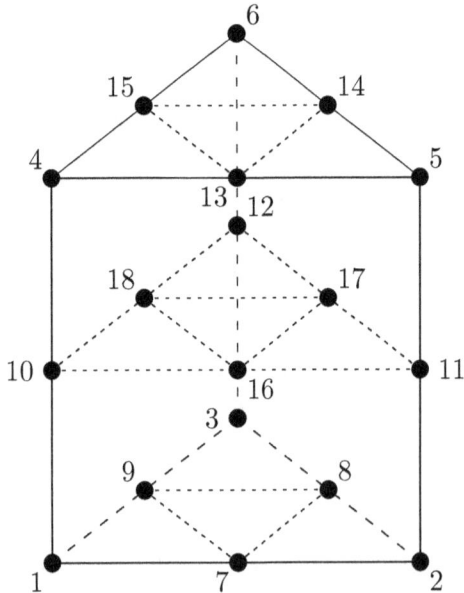

Edge Definition

Oriented edges	Corner nodes	Mid-node
E1	N1,N2	N7
E2	N2,N3	N8
E3	N3,N1	N9
E4	N1,N4	N10
E5	N2,N5	N11
E6	N3,N6	N12
E7	N4,N5	N13
E8	N5,N6	N14
E9	N6,N4	N15

Face Definition

Face	Corner nodes	Mid-edge nodes	Mid-face node	Oriented edges
F1	N1,N2,N5,N4	N7, N11,N13,N10	N16	E1, E5,-E7,-E4
F2	N2,N3,N6,N5	N8, N12,N14,N11	N17	E2, E6,-E8,-E5
F3	N3,N1,N4,N6	N9, N10,N15,N12	N18	E3, E4,-E9,-E6
F4	N1,N3,N2	N9, N8, N7		-E3,-E2,-E1
F5	N4,N5,N6	N13,N14,N15		E7, E8, E9

Notes

N1,...,N18   Grid point identification number. Integer ≥ 0 or blank, and $N1 \neq N2 \neq \ldots \neq N18$. Grid points N1...N3 are in consecutive order about one trilateral face. Grid points N4...N6 are in order in the same direction around the opposite trilateral face.

E1,...,E9   Edge identification number. The edges are oriented from the first to the second node. A negative edge (e.g., -E1) means that the edge is used in its reverse direction.

F1,...,F5   Face identification number. The faces are oriented so that the cross product of a vector from its first to second node, with a vector from its first to third node, is oriented outward.

3.3.3.4 Hexahedral Elements

CGNS supports three types of hexahedral elements, HEXA_8, HEXA_20, and HEXA_27. HEXA_8 elements are composed of eight nodes located at the eight geometric corners of the hexahedron. In addition, HEXA_20 and HEXA_27 elements have a node at the middle of each of the twelve edges; HEXA_27 adds a node at the middle of each of the six faces, and one at the cell center.

HEXA_8

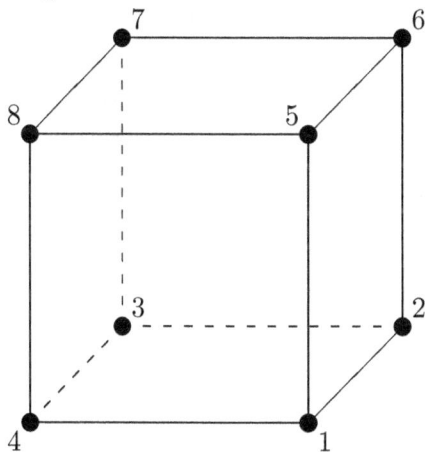

HEXA_20

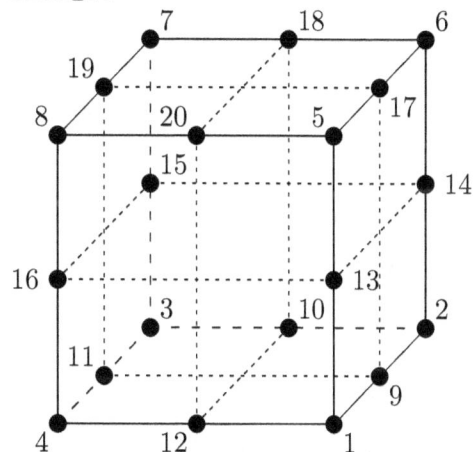

HEXA_27

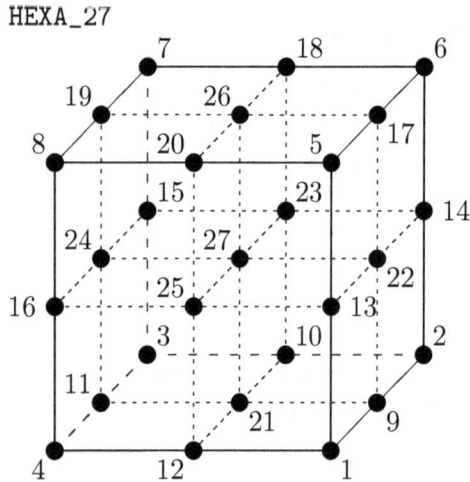

Edge Definition

Oriented edges	Corner nodes	Mid-node
E1	N1,N2	N9
E2	N2,N3	N10
E3	N3,N4	N11
E4	N4,N1	N12
E5	N1,N5	N13
E6	N2,N6	N14
E7	N3,N7	N15
E8	N4,N8	N16
E9	N5,N6	N17
E10	N6,N7	N18
E11	N7,N8	N19
E12	N8,N5	N20

Face Definition

Face	Corner nodes	Mid-edge nodes	Mid-face node	Oriented edges
F1	N1,N4,N3,N2	N12,N11,N10,N9	N21	-E4,-E3, -E2, -E1
F2	N1,N2,N6,N5	N9, N14,N17,N13	N22	E1, E6, -E9, -E5
F3	N2,N3,N7,N6	N10,N15,N18,N14	N23	E2, E7, -E10,-E6
F4	N3,N4,N8,N7	N11,N16,N19,N15	N24	E3, E8, -E11,-E7
F5	N1,N5,N8,N4	N13,N20,N16,N12	N25	E5,-E12,-E8, E4
F6	N5,N6,N7,N8	N17,N18,N19,N20	N26	E9, E10, E11, E12

Notes

N1,...,N27   Grid point identification number. Integer ≥ 0 or blank, and $N1 \neq N2 \neq \ldots \neq N27$. Grid points N1...N4 are in consecutive order about one quadrilateral face. Grid points N5...N8 are in order in the same direction around the opposite quadrilateral face.

E1,...,E12   Edge identification number. The edges are oriented from the first to the second node. A negative edge (e.g., -E1) means that the edge is used in its reverse direction.

F1,...,F6   Face identification number. The faces are oriented so that the cross product of a vector from its first to second node, with a vector from its first to third node, is oriented outward.

3.3.4   Unstructured Grid Example

Consider an unstructured zone in the shape of a cube, with each edge of the zone having three nodes. The resulting unstructured grid has a total of 27 nodes, as illustrated in the exploded figure below.

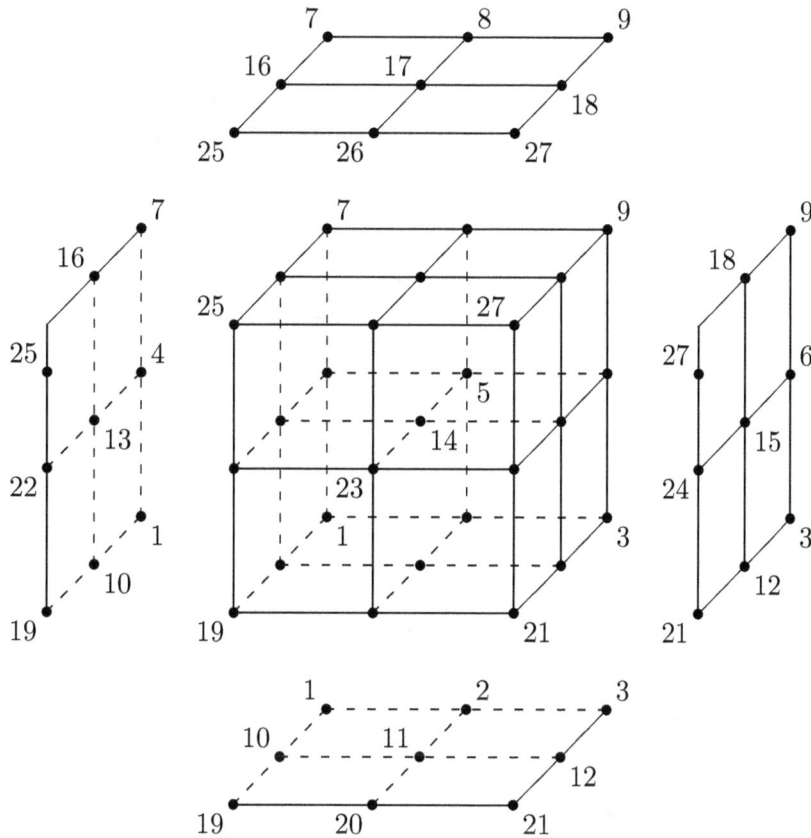

This zone contains eight hexahedral cells, numbered 1 to 8, and the cell connectivity is:

Element No.	Element Connectivity
1	1, 2, 5, 4, 10, 11, 14, 13
2	2, 3, 6, 5, 11, 12, 15, 14
3	4, 5, 8, 7, 13, 14, 17, 16
4	5, 6, 9, 8, 14, 15, 18, 17
5	10, 11, 14, 13, 19, 20, 23, 22
6	11, 12, 15, 14, 20, 21, 24, 23
7	13, 14, 17, 16, 22, 23, 26, 25
8	14, 15, 18, 17, 23, 24, 27, 26

In addition to the cells, the boundary faces could also be added to the element definition of this unstructured zone. There are 24 boundary faces in this zone, corresponding to element numbers 9 to 32. Each boundary face is of type QUAD_4. The table below shows the element connectivity of each boundary face, as well as the element number and face number of its parent cell.

Face	Element No.	Element Connectivity	Parent Cell	Parent Face
Left	9	1, 10, 13, 4	1	5
	10	4, 13, 16, 7	3	5
	11	10, 19, 22, 13	5	5
	12	13, 22, 25, 16	7	5
Right	13	3, 6, 15, 12	2	3
	14	6, 9, 18, 15	4	3
	15	12, 15, 24, 21	6	3
	16	15, 18, 27, 24	8	3
Bottom	17	1, 2, 11, 10	1	2
	18	2, 3, 12, 11	2	2
	19	10, 11, 20, 19	5	2
	20	11, 12, 21, 20	6	2
Top	21	7, 16, 17, 8	3	4
	22	8, 17, 18, 9	4	4
	23	16, 25, 26, 17	7	4
	24	17, 26, 27, 18	8	4
Back	25	1, 4, 5, 2	1	1
	26	2, 5, 6, 3	2	1
	27	4, 7, 8, 5	3	1
	28	5, 8, 9, 6	4	1
Front	29	19, 20, 23, 22	5	6
	30	20, 21, 24, 23	6	6
	31	22, 23, 26, 25	7	6
	32	23, 24, 27, 26	8	6

3.4 Multizone Interfaces

Figure 2 depicts three types of multizone interfaces, shown for structured zones. The first type is a 1-to-1 abutting interface, also referred to as matching or C0 continuous. The interface is a plane of vertices that are physically coincident between the adjacent zones. For structured zones, grid-coordinate lines perpendicular to the interface are continuous from one zone to the next; in 3-D, a 1-to-1 abutting interface is usually a logically rectangular region.

The second type of interface is mismatched abutting, where two zones touch but do not overlap (except for vertices and cell faces on the grid plane of the interface). Vertices on the interface may not be physically coincident between the two zones. Figure 2b identifies the vertices and face

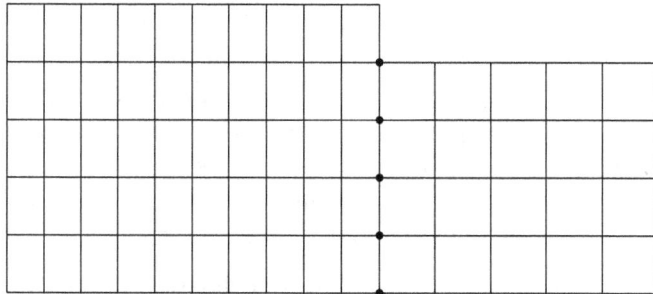

(a) 1-to-1 Abutting Interface

• Left-zone vertices on interface

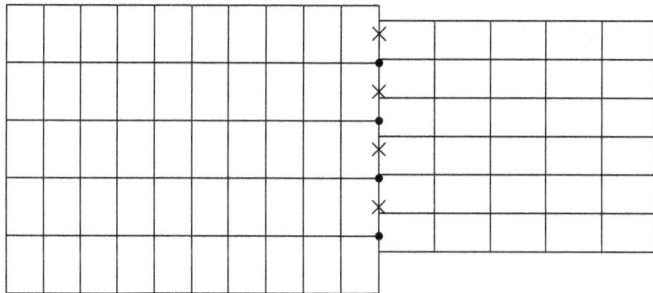

(b) Mismatched Abutting Interface

• Left-zone vertices on interface
× Left-zone face-centers on interface

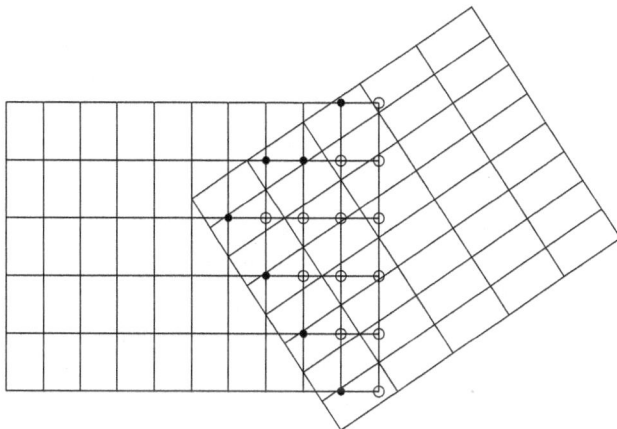

(c) Overset Interface

• Left-zone fringe points (vertices)
○ Left-zone overset-hole points (vertices)

Figure 2: Structured-Grid Multizone Interface Types

centers of the left zone that lay on the interface. Even for structured zones in 3-D, the vertices of a zone that constitute an interface patch may not form a logically rectangular region.

The third type of multizone interface is called overset and occurs when two zones overlap; in 3-D, the overlap is a 3-D region. For overset interfaces, one of the two zones takes precedence over the other; this establishes which solution in the overlap region to retain and which to discard. The region in a given zone where the solution is discarded is called an overset hole and the grid points outlining the hole are called fringe points. Figure 2c depicts an overlap region between two zones. The right zone takes precedence over the left zone, and the points identified in the figure are the fringe points and overset-hole points for the left zone. In addition, for the zone taking precedence, any bounding points (i.e. vertices on the bounding faces) of the zone that lay within the overlap region must also be identified.

Overset interfaces may also include multiple layers of fringe points outlining holes and at zone boundaries.

For the mismatched abutting and overset interfaces in Figure 2, the left zone plays the role of receiver zone and the right plays the role of donor zone.

4 Building-Block Structure Definitions

This section defines and describes low-level structures types that are used in the definition of more complex structures within the hierarchy.

4.1 Definition: `DataClass_t`

`DataClass_t` is an enumeration type that identifies the class of a given piece of data.

```
DataClass_t := Enumeration(
  Null,
  Dimensional,
  NormalizedByDimensional,
  NormalizedByUnknownDimensional,
  NondimensionalParameter,
  DimensionlessConstant,
  UserDefined ) ;
```

These classes divide data into different categories depending on dimensional units or normalization associated with the data. `Dimensional` specifies dimensional data. `NormalizedByDimensional` specifies nondimensional data that is normalized by dimensional reference quantities. In contrast, `NormalizedByUnknownDimensional` specifies nondimensional data typically found in completely nondimensional databases, where all field and reference data is nondimensional. `Nondimensional-Parameter` specifies nondimensional parameters such as Mach number and lift coefficient. Constants such as π are designated by `DimensionlessConstant`. The distinction between these different classes is further discussed in Section 5.

4.2 Definition: `Descriptor_t`

`Descriptor_t` is a documentation or annotation structure which contains a character string. Characters allowed within the string include newlines, tabs and other special characters; this potentially allows for unlimited documentation inclusion within the database. For example, a single `Descriptor_t` structure could be used to 'swallow' an entire ASCII file. In the hierarchical structures defined in the next sections, each allows for the inclusion of multiple `Descriptor_t` substructures. Conventions could be made for names of often-used `Descriptor_t` structure entities, such as ReadMe or YouReallyWantToReadMeFirst.

```
Descriptor_t :=
  {
  Data(char, 1, string_length) ;                              (r)
  } ;
```

where `string_length` is the length of the character string.

4.3 Definition: DimensionalUnits_t

DimensionalUnits_t describes the system of units used to measure dimensional data. It is composed of a set of enumeration types that define the mass, length, time, temperature and angle units.

```
MassUnits_t          := Enumeration( Null, Kilogram, Gram, Slug, PoundMass,
                                     UserDefined ) ;

LengthUnits_t        := Enumeration( Null, Meter, Centimeter, Millimeter,
                                     Foot, Inch, UserDefined ) ;

TimeUnits_t          := Enumeration( Null, Second, UserDefined ) ;

TemperatureUnits_t := Enumeration( Null, Kelvin, Celsius, Rankine,
                                     Fahrenheit, UserDefined ) ;

AngleUnits_t         := Enumeration( Null, Degree, Radian, UserDefined ) ;

DimensionalUnits_t :=
  {
  MassUnits_t          MassUnits ;                                          (r)
  LengthUnits_t        LengthUnits ;                                        (r)
  TimeUnits_t          TimeUnits ;                                          (r)
  TemperatureUnits_t TemperatureUnits ;                                     (r)
  AngleUnits_t         AngleUnits ;                                         (r)
  } ;
```

System International (SI) units have MassUnits = Kilogram; LengthUnits = Meter; TimeUnits = Second; TemperatureUnits = Kelvin; and AngleUnits = Radian.

For an entity of type DimensionalUnits_t, if all the elements of that entity have the value Null (i.e. MassUnits = Null, etc.), this is equivalent to stating that the data described by the entity is nondimensional.

4.4 Definition: DimensionalExponents_t

DimensionalExponents_t describes the dimensionality of data by defining the exponents associated with each of the fundamental units.

```
DimensionalExponents_t :=
  {
  real MassExponent ;                                                       (r)
  real LengthExponent ;                                                     (r)
  real TimeExponent ;                                                       (r)
```

```
real TemperatureExponent ;                              (r)
real AngleExponent ;                                    (r)
} ;
```

For example, an instance of DimensionalExponents_t that describes velocity is,

```
DimensionalExponents_t =
  {{
  MassExponent          =  0 ;
  LengthExponent        = +1 ;
  TimeExponent          = -1 ;
  TemperatureExponent   =  0 ;
  AngleExponent         =  0 ;
  }} ;
```

4.5 Definition: GridLocation_t

GridLocation_t identifies locations with respect to the grid; it is an enumeration type.

```
GridLocation_t := Enumeration(
  Null,
  Vertex,
  CellCenter,
  FaceCenter,
  IFaceCenter,
  JFaceCenter,
  KFaceCenter,
  EdgeCenter,
  UserDefined ) ;
```

Vertex is coincident with the grid vertices. CellCenter is the center of a cell; this is also appropriate for entities associated with cells but not necessarily with a given location in a cell. For structured zones, IFaceCenter is the center of a face in 3-D whose computational normal points in the i direction. JFaceCenter and KFaceCenter are similarly defined, again only for structured zones. FaceCenter is the center of a generic face which can point in any coordinate direction. These are also appropriate for entities associated with a face, but not located at a specific place on the face. EdgeCenter is the center of an edge. See Section 3.2 for descriptions of cells, faces and edges.

All of the entities of type GridLocation_t defined in this document use a default value of Vertex.

4.6 Definition: IndexArray_t

IndexArray_t specifies an array of indices. An argument is included that allows for specifying the data type of each index; typically the data type will be integer (int). IndexArray_t defines an array of indices of size ArraySize, where the dimension of each index is IndexDimension.

```
IndexArray_t< int IndexDimension, int ArraySize, DataType > :=
  {
  Data( DataType, 2, [IndexDimension,ArraySize] ) ;                    (r)
  } ;
```

4.7 Definition: IndexRange_t

IndexRange_t specifies the beginning and ending indices of a subrange. The subrange may describe a portion of a grid line, grid plane, or grid volume.

```
IndexRange_t< int IndexDimension > :=
  {
  int[IndexDimension] Begin ;                                          (r)
  int[IndexDimension] End ;                                            (r)
  } ;
```

where Begin and End are the indices of the opposing corners of the subrange.

4.8 Definition: Rind_t

Rind_t describes the number of rind planes associated with a data array containing grid coordinates, flow-solution data or any other grid-related discrete data for structured zones.

```
Rind_t< int IndexDimension > :=
  {
  int[2*IndexDimension] RindPlanes ;                                   (r)
  } ;
```

RindPlanes contains the number of rind planes attached to the minimum and maximum faces of a zone. The face corresponding to each index n of RindPlanes in 3-D is:

$$
\begin{array}{ll}
n = 1 \rightarrow & i\text{-min} \qquad\qquad n = 2 \rightarrow \quad i\text{-max} \\
n = 3 \rightarrow & j\text{-min} \qquad\qquad n = 4 \rightarrow \quad j\text{-max} \\
n = 5 \rightarrow & k\text{-min} \qquad\qquad n = 6 \rightarrow \quad k\text{-max}
\end{array}
$$

For a 3-D grid whose 'core' size is II×JJ×KK, a value of RindPlanes = [a,b,c,d,e,f] indicates that the range of indices for the grid with rind is:

$$
\begin{array}{ll}
i: & (1 - a, \ II + b) \\
j: & (1 - c, \ JJ + d) \\
k: & (1 - e, \ KK + f)
\end{array}
$$

5 Data-Array Structure Definitions

This section defines the structure type `DataArray_t` for describing data arrays. This general-purpose structure is used to declare data arrays and scalars throughout the CGNS hierarchy. It is used to describe grid coordinates, flow-solution data, governing flow parameters, boundary-condition data, and other information. For most of these different types of CFD data, we have also established a list of standardized identifiers for entities of type `DataArray_t`. For example, `Density` is used for data arrays containing static density. The list of standardized data-name identifiers is provided in Annex A.

We address five classes of data with the `DataArray_t` structure type:

(a) dimensional data (e.g. velocity in units of m/s);

(b) nondimensional data normalized by dimensional reference quantities;

(c) nondimensional data with associated nondimensional reference quantities;

(d) nondimensional parameters (e.g. Reynolds number, pressure coefficient);

(e) pure constants (e.g. π, e).

The first two of these classes often occur within the same test case, where each piece of data is either dimensional itself or normalized by a dimensional quantity. The third data class is typical of a completely nondimensional test case, where all field data and reference quantities are nondimensional. The forth class, nondimensional parameters, are universal in CFD, although not always consistently defined. The individual components of nondimensional parameters may be data from any of the first three classes.

Each of the five classes of data requires different information to describe dimensional units or normalization associated with the data. These requirements are reflected in the structure definition for `DataArray_t`.

The remainder of this section is as follows: the structure type `DataArray_t` is first defined. Then the class identification and data manipulation is discussed in Section 5.2 for each of the five data classes. Finally, examples of `DataArray_t` entities are presented in Section 5.3.

5.1 Definition: `DataArray_t`

`DataArray_t` describes a multi-dimensional data array of given type, dimensionality and size in each dimension. The data may be dimensional, nondimensional or pure constants. Qualifiers are provided to describe dimensional units or normalization information associated with the data.

```
DataArray_t< DataType, int Dimension, int[Dimension] DimensionValues > :=
  {
  List( Descriptor_t Descriptor1 ... DescriptorN ) ;                    (o)

  Data( DataType, Dimension, DimensionValues ) ;                        (r)
```

```
DataClass_t DataClass ;                                        (o)

DimensionalUnits_t DimensionalUnits ;                          (o)

DimensionalExponents_t DimensionalExponents ;                  (o)

DataConversion_t DataConversion ;                             (o)
} ;
```

Notes

1. Default names for the `Descriptor_t` list are as shown; users may choose other legitimate names. Legitimate names must be unique within a given instance of `DataArray_t` and shall not include the names `DataClass`, `DimensionalUnits`, `DimensionalExponents`, or `DataConversion`.
2. `Data()` is the only required field for `DataArray_t`.

`DataArray_t` requires three structure parameters: `Dimension` is the dimensionality of the data array; `DimensionValues` is an array of length `Dimension` that contains the size of the data arrays in each dimension; and `DataType` is the data type of the data stored. `DataType` will usually be `real`, but other data types are permissible.

The optional entities `DataClass`, `DimensionalUnits`, `DimensionalExponents` and `DataConversion` provide information on dimensional units and normalization associated with the data. The function of these qualifiers is provided in the next section.

This structure type is formulated to describe an array of scalars. Therefore, for vector quantities (e.g. the position vector or the velocity vector), separate structure entities are required for each component of the vector. For example, the cartesian coordinates of a 3-D grid are described by three separate `DataArray_t` entities: one for x, one for y and one for z (see Example 7-A).

5.1.1 Definition: `DataConversion_t`

`DataConversion_t` contains conversion factors for recovering raw dimensional data from given nondimensional data. These conversion factors are typically associated with nondimensional data that is normalized by dimensional reference quantities.

```
DataConversion_t :=
  {
  real ConversionScale ;                                       (r)

  real ConversionOffset ;                                      (r)
  } ;
```

Given a nondimensional piece of data, `Data(nondimensional)`, the conversion to 'raw' dimensional form is:

```
Data(raw) = Data(nondimensional)*ConversionScale + ConversionOffset
```

These conversion factors are further described in Section 5.2.2.

5.2   Data Manipulation

The optional entities of `DataArray_t` provide information for manipulating the data, including changing units or normalization. This section describes the rules under which these optional entities operate and the specific manipulations that can be performed on the data.

Within a given instance of `DataArray_t`, the class of data and all information required for manipulations may be completely and precisely specified by the entities `DataClass`, `DimensionalUnits`, `DimensionalExponents` and `DataConversion`. `DataClass` identifies the class of data and governs the manipulations that can be performed. Each of the five data classes is treated separately in the subsequent sections.

The entities `DataClass` and `DimensionalUnits` serve special functions in the CGNS hierarchy. If `DataClass` is absent from a given instance of `DataArray_t`, then its value is determined from 'global' data. This global data may be set at any level of the CGNS hierarchy with the data set at the lowest level taking precedence. `DimensionalUnits` may be similarly set by global data. The rules for determining the appropriate set of global data to apply is further detailed in Section 6.3.

This alternate functionality provides a measure of economy in describing dimensional units or normalization within the hierarchy. Examples that make use of global data are presented in Section 7.2 and Section 7.6 for grid and flow solution data. The complete two-zone case of Annex B also depicts this alternate functionality.

5.2.1   Dimensional Data

If `DataClass = Dimensional`, the data is dimensional. The optional qualifiers `DimensionalUnits` and `DimensionalExponents` describe dimensional units associated with the data. These qualifiers are provided to specify the system of dimensional units and the dimensional exponents, respectively. For example, if the data is the x-component of velocity, then `DimensionalUnits` will state that the pertinent dimensional units are, say, `Meter` and `Second`; `DimensionalExponents` will specify the pertinent dimensional exponents are `LengthExponent = 1` and `TimeExponent = -1`. Combining the information gives the units m/s. Examples showing the use of these two qualifiers are provided in Section 5.3.

If `DimensionalUnits` is absent, then the appropriate set of dimensional units is obtained from 'global' data. The rules for determining this appropriate set of 'global' dimensional units are presented in Section 6.3.

If `DimensionalExponents` is absent, then the appropriate dimensional exponents can be determined by convention if the specific instance of `DataArray_t` corresponds to one of the standardized data-name identifiers listed in Annex A. Otherwise, the exponents are unspecified. We strongly recommend inclusion of the `DimensionalExponents` qualifier whenever the data is dimensional and the instance of `DataArray_t` is not among the list of standardized identifiers.

5.2.2 Nondimensional Data Normalized by Dimensional Quantities

If `DataClass = NormalizedByDimensional`, the data is nondimensional and is normalized by dimensional reference quantities. All optional entities in `DataArray_t` are used. `DataConversion` contains factors to convert the nondimensional data to 'raw' dimensional data; these factors are `ConversionScale` and `ConversionOffset`. The conversion process is as follows:

$$\text{Data(raw)} = \text{Data(nondimensional)} * \text{ConversionScale} + \text{ConversionOffset}$$

where `Data(nondimensional)` is the original nondimensional data, and `Data(raw)` is the converted raw data. This converted raw data is dimensional, and the optional qualifiers `DimensionalUnits` and `DimensionalExponents` describe the appropriate dimensional units and exponents. Note that `DimensionalUnits` and `DimensionalExponents` also describe the units for `ConversionScale` and `ConversionOffset`.

If `DataConversion` is absent, the equivalent defaults are `ConversionScale = 1` and `ConversionOffset = 0`. If either `DimensionalUnits` or `DimensionalExponents` is absent, follow the rules described in the previous section.

Note that functionally there is little difference between these first two data classes (`DataClass = Dimensional` and `NormalizedByDimensional`). In the first case the data is dimensional, and in the second, the converted raw data is dimensional. Also, the equivalent defaults for `DataConversion` produce no changes in the data; hence, it is almost the same as stating the original data is dimensional.

5.2.3 Nondimensional Data Normalized by Unknown Dimensional Quantities

If `DataClass = NormalizedByUnknownDimensional`, the data is nondimensional and is normalized by some unspecified dimensional quantities. This type of data is typical of a completely nondimensional test case, where all field data and all reference quantities are nondimensional.

Only the `DimensionalExponents` qualifier is used in this case, although it is expected that this qualifier will be seldom utilized in practice. For entities of `DataArray_t` that are not among the list of standardized data-name identifiers, the qualifier could provide useful information by defining the exponents of the dimensional form of the nondimensional data.

Rather than providing qualifiers to describe the normalization of the data, we instead dictate that all data of type `NormalizedByUnknownDimensional` in a given database be nondimensionalized consistently. This is done by picking one set of mass, length, time and temperature scales and normalizing all appropriate data by these scales. We describe this process in detail in the following. Annex B also shows a completely nondimensional database where consistent normalization is used throughout.

The practice of nondimensionalization within flow solvers and other application codes is quite popular. The problem with this practice is that to manipulate the data from a given code, one must often know the particulars of the nondimensionalization used. This largely results from what we call inconsistent normalization—more than the minimum required scales are used to normalize data within the code. For example, in the OVERFLOW flow solver, the following nondimensionalization is used:

$$\tilde{x} = x/L, \quad \tilde{u} = u/c_\infty, \quad \tilde{\rho} = \rho/\rho_\infty,$$
$$\tilde{y} = y/L, \quad \tilde{v} = v/c_\infty, \quad \tilde{p} = p/(\rho_\infty c_\infty^2),$$
$$\tilde{z} = z/L, \quad \tilde{w} = w/c_\infty, \quad \tilde{\mu} = \mu/\mu_\infty,$$

where (x, y, z) are the cartesian coordinates, (u, v, w) are the cartesian components of velocity, ρ is static density, p is static pressure, c is the static speed of sound, and μ is the molecular viscosity. In this example, tilde quantities ($\tilde{\ }$) are nondimensional and all others are dimensional. Four dimensional scales are used for normalization: L (a unit length), ρ_∞, c_∞ and μ_∞. However, only three fundamental dimensional units are represented: mass, length and time. The extra normalizing scale leads to inconsistent normalization. The primary consequence of this is additional nondimensional parameters, such as Reynolds number, appearing in the nondimensionalized governing equations where none are found in the original dimensional equations. Many definitions, including skin friction coefficient, also have extra terms appearing in the nondimensionalized form. This adds unnecessary complication to any data or equation manipulation associated with the flow solver.

Consistent normalization avoids many of these problems. Here the number of scales used for normalization is the same as the number of fundamental dimensional units represented by the data. Using consistent normalization, the resulting nondimensionalized form of equations and definitions is identical to their original dimensional formulations. One piece of evidence to support this assertion is that it is not possible to form any nondimensional parameters from the set of dimensional scales used for normalization.

An important fallout of consistent normalization is that the actual scales used for normalization become immaterial for all data manipulation processes. To illustrate this consider the following nondimensionalization procedure: let M (mass), L (length) and T (time) be arbitrary dimensional scales by which all data is normalized (neglect temperature data for the present). The nondimensional data follows:

$$x' = x/L, \quad u' = u/(L/T), \quad \rho' = \rho/(M/L^3),$$
$$y' = y/L, \quad v' = v/(L/T), \quad p' = p/(M/(LT^2)),$$
$$z' = z/L, \quad w' = w/(L/T), \quad \mu' = \mu/(M/(LT)),$$

where primed quantities are nondimensional and all others are dimensional.

Consider an existing database where all field data and all reference data is nondimensional and normalized as shown. Assume the database has a single reference state given by,

$$x'_{\text{ref}} = x_{\text{ref}}/L, \quad u'_{\text{ref}} = u_{\text{ref}}/(L/T), \quad \rho'_{\text{ref}} = \rho_{\text{ref}}/(M/L^3),$$
$$y'_{\text{ref}} = y_{\text{ref}}/L, \quad v'_{\text{ref}} = v_{\text{ref}}/(L/T), \quad p'_{\text{ref}} = p_{\text{ref}}/(M/(LT^2))$$
$$z'_{\text{ref}} = z_{\text{ref}}/L, \quad w'_{\text{ref}} = w_{\text{ref}}/(L/T), \quad \mu'_{\text{ref}} = \mu_{\text{ref}}/(M/(LT)).$$

If a user wanted to change the nondimensionalization of grid-point pressures, the procedure is straightforward. Let the desired new normalization be given by $p''_{ijk} = p_{ijk}/(\rho_{\text{ref}} c_{\text{ref}}^2)$, where all terms on the right-hand-side are *dimensional*, and as such they are unknown to the database user. However, the desired manipulation is possible using only nondimensional data provided in the database,

$$
\begin{aligned}
p''_{ijk} &\equiv p_{ijk}/(\rho_{\mathrm{ref}}c_{\mathrm{ref}}^2) \\
&= \frac{p_{ijk}}{M/(LT^2)}\frac{M/L^3}{\rho_{\mathrm{ref}}}\left[\frac{L/T}{c_{\mathrm{ref}}}\right]^2 \\
&= p'_{ijk}/(\rho'_{\mathrm{ref}}(c'_{\mathrm{ref}})^2)
\end{aligned}
$$

Thus, the desired renormalization is possible using the database's nondimensional data as if it were actually dimensional. There is, in fact, a high degree of equivalence between dimensional data and consistently normalized nondimensional data. The procedure shown in this example should extend to any desired renormalization, provided the needed reference-state quantities are included in the database.

This example points out two stipulations that we now dictate for data in the class `NormalizedBy-UnknownDimensional`,

(a) All nondimensional data within a given database that has `DataClass = NormalizedBy-UnknownDimensional` shall be consistently normalized.

(b) Any nondimensional reference state appearing in a database should be sufficiently populated with reference quantities to allow for renormalization procedures.

The second of these stipulations is somewhat ambiguous, but good practice would suggest that a flow solver, for example, should output to the database enough static and/or stagnation reference quantities to sufficiently define the state.

Annex B shows an example of a well-populated reference state.

With these two stipulations, we contend the following:

- The dimensional scales used to nondimensionalize all data are immaterial, and there is no need to identify these quantities in a CGNS database.

- The dimensional scales need not be reference-state quantities provided in the database. For example, a given database could contain freestream reference state conditions, but all the data is normalized by sonic conditions (which are not provided in the database).

- All renormalization procedures can be carried out treating the data as if it were dimensional with a consistent set of units.

- Any application code that internally uses consistent normalization can use the data provided in a CGNS database without modification or transformation to the code's internal normalization.

Before ending this section, we note that the OVERFLOW flow solver mentioned above (or any other application code that internally uses inconsistent normalization) could easily read and write data to a nondimensional CGNS database that conforms to the above stipulations. On output, the code could renormalize data so it is consistently normalized. Probably, the easiest method would be to remove the molecular viscosity scale (μ_∞), and only use L, ρ_∞ and c_∞ for all normalizations (recall these are dimensional scales). The only change from the above example would be the nondimensionalization of viscosity, which would become, $\tilde{\mu} = \mu/(\rho_\infty c_\infty L)$. The code could then output all field data as,

$$\tilde{x}_{ijk} = x_{ijk}/L, \qquad \tilde{u}_{ijk} = u_{ijk}/c_\infty, \qquad \tilde{\rho}_{ijk} = \rho_{ijk}/\rho_\infty,$$
$$\tilde{y}_{ijk} = y_{ijk}/L, \qquad \tilde{v}_{ijk} = v_{ijk}/c_\infty, \qquad \tilde{p}_{ijk} = p_{ijk}/(\rho_\infty c_\infty^2),$$
$$\tilde{z}_{ijk} = z_{ijk}/L, \qquad \tilde{w}_{ijk} = w_{ijk}/c_\infty, \qquad \tilde{\tilde{\mu}}_{ijk} = \mu_{ijk}/(\rho_\infty c_\infty L),$$

and output the freestream reference quantities,

$$\tilde{u}_\infty = u_\infty/c_\infty, \qquad\qquad \tilde{\rho}_\infty = \rho_\infty/\rho_\infty = 1,$$
$$\tilde{v}_\infty = v_\infty/c_\infty, \qquad\qquad \tilde{p}_\infty = p_\infty/(\rho_\infty c_\infty^2) = 1/\gamma,$$
$$\tilde{w}_\infty = w_\infty/c_\infty, \qquad\qquad \tilde{\tilde{\mu}}_\infty = \mu_\infty/(\rho_\infty c_\infty L) \sim O(1/Re),$$
$$\tilde{c}_\infty = c_\infty/c_\infty = 1, \qquad\qquad \tilde{L} = L/L = 1,$$

where γ is the specific heat ratio (assumes a perfect gas) and Re is the Reynolds number.

On input, the flow solver should be able to recover its internal normalizations from the data in a nondimensional CGNS database by treating the data as if it were dimensional.

5.2.4 Nondimensional Parameters

If `DataClass = NondimensionalParameter`, the data is a nondimensional parameter (or array of nondimensional parameters). Examples include Mach number, Reynolds number and pressure coefficient. These parameters are prevalent in CFD, although their definitions tend to vary between different application codes. A list of standardized data-name identifiers for nondimensional parameters is provided in Annex A.4.

We distinguish nondimensional parameters from other data classes by the fact that they are *always* dimensionless. In a completely nondimensional database, they are distinct in that their normalization is not necessarily consistent with other data.

Typically, the `DimensionalUnits`, `DimensionalExponents` and `DataConversion` qualifiers are not used for nondimensional parameters; although, there are a few situations where they may be used (these are discussed below). Rather than rely on optional qualifiers to describe the normalization, we establish the convention that *any nondimensional parameters should be accompanied by their defining scales*; this is further discussed in Annex A.4. An example is Reynolds number defined as $Re = VL_R/\nu$, where V, L_R and ν are velocity, length, and viscosity scales, respectively. Note that these defining scales may be dimensional or nondimensional data. We establish the data-name identifiers `Reynolds`, `Reynolds_Velocity`, `Reynolds_Length` and `Reynolds_ViscosityKinematic` for the Reynolds number and its defining scales. Anywhere an instance of `DataArray_t` is found with the identifier `Reynolds`, there should also be entities for the defining scales. An example of this use for Reynolds number is given in Section 5.3.

In certain situations, it may be more convenient to use the optional qualifiers of `DataArray_t` to describe the normalization used in nondimensional parameters. These situations must satisfy two requirements: First, the defining scales are dimensional; and second, the nondimensional parameter is a normalization of a single 'raw' data quantity and it is clear what this raw data is. Examples that satisfy this second constraint are pressure coefficient, where the raw data is static pressure, and lift coefficient, where the raw data is the lift force. Conversely, Reynolds number is a parameter that violates the second requirement—there are three pieces of raw data rather than one that make up Re. For nondimensional parameters that satisfy these two requirements, the qualifiers `DimensionalUnits`, `DimensionalExponents` and `DataConversion` may be used as in Section 5.2.2 to recover the raw dimensional data.

5.2.5 Dimensionless Constants

If `DataClass = DimensionlessConstant`, the data is a constant (or array of constants) with no associated dimensional units. The `DimensionalUnits`, `DimensionalExponents` and `DataConversion` qualifiers are not used.

5.3 Data-Array Examples

This section presents five examples of data-array entities and illustrates the use of optional information for describing dimensional and nondimensional data.

Example 5-A: One-Dimensional Data Array, Constants

A one-dimensional array of integers; the array is the integers from 1 to 10. The data is pure constants.

```
!  DataType = int
!  Dimension = 1
!  DimensionValues = 10
DataArray_t<int, 1, 10> Data1 =
  {{
  Data(int, 1, 10) = [1, 2, 3, 4, 5, 6, 7, 8, 9, 10] ;

  DataClass_t DataClass = DimensionlessConstant ;
  }} ;
```

The structure parameters for `DataArray_t` state the data is an one-dimensional integer array of length ten. The value of `DataClass` indicates the data is unitless constants.

Example 5-B: Two-Dimensional Data Array, Pressures

A two-dimensional array of pressures with size 11×9 given by the array `P(i,j)`. The data is dimensional with units of N/m^2 (i.e., $kg/(m\text{-}s^2)$). Note that `Pressure` is the data-name identifier for static pressure.

```
!  DataType = real
!  Dimension = 2
!  DimensionValues = [11,9]
DataArray_t<real, 2, [11,9]> Pressure =
```

```
{{
Data(real, 2, [11,9]) = ((P(i,j), i=1,11), j=1,9) ;

DataClass_t DataClass = Dimensional ;

DimensionalUnits_t DimensionalUnits =
  {{
  MassUnits        = Kilogram ;
  LengthUnits      = Meter ;
  TimeUnits        = Second ;
  TemperatureUnits = Null ;
  AngleUnits       = Null ;
  }} ;

DimensionalExponents_t DimensionalExponents =
  {{
  MassExponent        = +1 ;
  LengthExponent      = -1 ;
  TimeExponent        = -2 ;
  TemperatureExponent =  0 ;
  AngleExponent       =  0 ;
  }} ;
}} ;
```

From the data-name identifier conventions presented in Annex A, `Pressure` has a floating-point data type; hence, the appropriate structure parameter for `DataArray_t` is `real`.

The value of `DataClass` indicates that the data is dimensional, and both the dimensional units and dimensional exponents are provided. `DimensionalUnits` specifies that the system of units is kg-m-s, and `DimensionalExponents` specified the appropriate exponents for pressure. Combining the information gives pressure as kg/(m-s^2). `DimensionalExponents` could have been defaulted, since the dimensional exponents are given in Annex A for the data-name identifier `Pressure`.

Note that FORTRAN multidimensional array indexing is used to store the data; this is reflected in the FORTRAN-like implied do-loops for `P(i,j)`.

Example 5-C: Three-Dimensional Data Array, Nondimensional Static Enthalpy

A 3-D array of size $33 \times 9 \times 17$ containing nondimensional static enthalpy. The data is normalized by freestream velocity as follows:

$$\bar{h}_{i,j,k} = \frac{h_{i,j,k}}{q_{\text{ref}}^2},$$

where $\bar{h}_{i,j,k}$ is nondimensional static enthalpy. The freestream velocity is dimensional with a value of 10 m/s.

```
!  DataType = real
!  Dimension = 3
```

```
!  DimensionValues = [33,9,17]
DataArray_t<real, 3, [33,9,17]> Enthalpy =
  {{
  Data(real, 3, [33,9,17]) = (((H(i,j,k), i=1,33), j=1,9), k=1,17) ;

  DataClass_t DataClass = NormalizedByDimensional ;

  DataConversion_t DataConversion =
    {{
    real ConversionScale  = 100 ;
    real ConversionOffset = 0 ;
    }} ;

  DimensionalUnits_t DimensionalUnits =
    {{
    MassUnits        = Null ;
    LengthUnits      = Meter ;
    TimeUnits        = Second ;
    TemperatureUnits = Null ;
    AngleUnits       = Null ;
    }} ;

  DimensionalExponents_t DimensionalExponents =
    {{
    MassExponent        =  0 ;
    LengthExponent      = +2 ;
    TimeExponent        = -2 ;
    TemperatureExponent =  0 ;
    AngleExponent       =  0 ;
    }} ;
  }} ;
```

From Annex A, the identifier for static enthalpy is `Enthalpy` and its data type is `real`.

The value of `DataClass` indicates that the data is nondimensional and normalized by a dimensional reference quantity. `DataConversion` provides the conversion factors for recovering the raw static enthalpy, which has units of m^2/s^2 as indicated by `DimensionalUnits` and `DimensionalExponents`. Note that `DimensionalExponents` could have been defaulted using the conventions for the data-name identifier `Enthalpy`.

Example 5-D: Three-Dimensional Data Array, Nondimensional Database

The previous example for nondimensional enthalpy is repeated for a completely nondimensional database.

```
!  DataType = real
!  Dimension = 3
```

```
!  DimensionValues = [33,9,17]
DataArray_t<real, 3, [33,9,17]> Enthalpy =
  {{
  Data(real, 3, [33,9,17]) = (((H(i,j,k), i=1,33), j=1,9), k=1,17) ;

  DataClass_t DataClass = NormalizedByUnknownDimensional ;
  }} ;
```

The value of `DataClass` indicates the appropriate class.

Example 5-E: Data Arrays for Reynolds Number

Reynolds number of 1.554×10^6 based on a velocity scale of 10 m/s, a length scale of 2.3 m and a kinematic viscosity scale of 1.48×10^{-5} m^2/s. Assume the database has globally set the dimensional units to kg-m-s and the global default data class to dimensional (`DataClass = Dimensional`).

```
!  DataType = real
!  Dimension = 1
!  DimensionValues = 1
DataArray_t<real, 1, 1> Reynolds =
  {{
  Data(real, 1, 1) = 1.554e+06 ;

  DataClass_t DataClass = NondimensionalParameter ;
  }} ;

DataArray_t<real, 1, 1> Reynolds_Velocity =
  {{
  Data(real, 1, 1) = 10. ;
  }} ;

DataArray_t<real, 1, 1> Reynolds_Length =
  {{
  Data(real, 1, 1) = 2.3 ;
  }} ;

DataArray_t<real, 1, 1> Reynolds_ViscosityKinematic =
  {{
  Data(real, 1, 1) = 1.48e-05 ;
  }} ;
```

`Reynolds` contains the value of the Reynolds number, and the value of its `DataClass` qualifier designates it as a nondimensional parameter. By conventions described in Annex A.4, the defining scales are contained in the associated entities `Reynolds_Velocity`, `Reynolds_Length`, and `Reynolds_ViscosityKinematic`. Since each of these entities contain no qualifiers, global information is used to decipher that they are all dimensional with *kg-m-s* units. The structure parameters for each `DataArray_t` entity state that they contain a real scalar.

If a user wanted to convey the dimensional units of the defining scales using optional qualifiers of `DataArray_t`, then the last three entities in this example would have a form similar to that in Example 5-B.

6 Hierarchical Structures

This section presents the structure-type definitions for the top levels of the CGNS hierarchy. As stated in Section 2, the hierarchy is topologically based, where the overall organization is by zones. All information pertaining to a given zone, including grid coordinates, flow solution, and other related data, is contained within that zone's structure entity. Figure 1 depicts this topologically based hierarchy. The CGNS database entry level structure type is defined in Section 6.1, and the zone structure is defined in Section 6.2. This section concludes with a discussion of globally applicable data.

6.1 CGNS Entry Level Structure Definition: CGNSBase_t

The highest level structure in a CGNS database is CGNSBase_t. It contains the cell dimension and physical dimension of the computational grid and lists of zones and families making up the domain. Globally applicable information, including a reference state, a set of flow equations, dimensional units, time step or iteration information, and convergence history are also attached. In addition, structures for describing or annotating the database are also provided; these same descriptive mechanisms are provided for structures at all levels of the hierarchy.

```
CGNSBase_t :=
  {
  List( Descriptor_t Descriptor1 ... DescriptorN ) ;                        (o)

  int CellDimension ;                                                       (r)
  int PhysicalDimension ;                                                   (r)

  BaseIterativeData_t BaseIterativeData ;                                   (o)

  List( Zone_t<CellDimension, PhysicalDimension> Zone1 ... ZoneN ) ;       (o)

  ReferenceState_t ReferenceState ;                                        (o)

  SimulationType_t SimulationType ;                                         (o)

  DataClass_t DataClass ;                                                   (o)

  DimensionalUnits_t DimensionalUnits ;                                     (o)

  FlowEquationSet_t<CellDimension> FlowEquationSet ;                        (o)

  ConvergenceHistory_t GlobalConvergenceHistory ;                          (o)

  List( IntegralData_t IntegralData1... IntegralDataN ) ;                   (o)
```

```
    List( Family_t Family1... FamilyN ) ;                                    (o)
    } ;
```

Notes

1. Default names for the `Descriptor_t`, `Zone_t`, `IntegralData_t`, and `Family_t` lists are as shown; users may choose other legitimate names. Legitimate names must be unique at this level and shall not include the names `BaseIterativeData`, `DataClass`, `DimensionalUnits`, `FlowEquationSet`, `GlobalConvergenceHistory`, `ReferenceState`, or `SimulationType`.
2. The number of entities of type `Zone_t` defines the number of zones in the domain.
3. `CellDimension` and `PhysicalDimension` are the only required fields. The `Descriptor_t`, `Zone_t` and `IntegralData_t` lists may be empty, and all other optional fields absent.

Note that we make the distinction between the following:

`IndexDimension`	Number of different indices required to reference a node (e.g., $1 = i$, $2 = i, j$, $3 = i, j, k$)
`CellDimension`	Dimensionality of the cell in the mesh (e.g., 3 for a volume cell, 2 for a face cell)
`PhysicalDimension`	Number of coordinates required to define a node position (e.g., 1 for 1-D, 2 for 2-D, 3 for 3-D)

These three dimensions may differ depending on the mesh. For example, an unstructured triangular surface mesh representing the wet surface of an aircraft will have:

- `IndexDimension` = 1 (always for unstructured)

- `CellDimension` = 2 (face elements)

- `PhysicalDimension` = 3 (needs x, y, z coordinates since it is a 3D surface)

For a structured zone, the quantities `IndexDimension` and `CellDimension` are always equal. For an unstructured zone, `IndexDimension` always equals 1. Therefore, storing `CellDimension` at the `CGNSBase_t` level will automatically define the `IndexDimension` value for each zone.

On the other hand we assume that all zones of the base have the same `CellDimension`, e.g., if `CellDimension` is 3, all zones must be composed of 3D cells within the `CGNSBase_t`.

We need `IndexDimension` for both structured and unstructured zones in order to use original data structures such as `GridCoordinates_t`, `FlowSolution_t`, `DiscreteData_t`, etc. `CellDimension` is necessary to express the interpolants in `ZoneConnectivity` with an unstructured zone (mismatch or overset connectivity). When the cells are bidimensional, two interpolants per node are required, while when the cells are tridimensional, three interpolants per node must be provided. `PhysicalDimension` becomes useful when expressing quantities such as the `InwardNormalList` in the `BC_t` data structure. It's possible to have a mesh where `IndexDimension` = 2 but the normal vectors still require x, y, z components in order to be properly defined. Consider, for example, a structured surface mesh in the 3D space.

Information about the number of time steps or iterations being recorded, and the time and/or iteration values at each step, may be contained in the `BaseIterativeData` structure.

Data specific to each zone in a multizone case is contained in the list of `Zone_t` structure entities.

Reference data applicable to the entire CGNS database is contained in the `ReferenceState` structure; quantities such as Reynolds number and freestream Mach number are contained here (for external flow problems).

`SimulationType` is an enumeration type identifying the type of simulation.

```
SimulationType_t := Enumeration (
  Null,
  UserDefined,
  TimeAccurate,
  NonTimeAccurate ) ;
```

`DataClass` describes the global default for the class of data contained in the CGNS database. If the CGNS database contains dimensional data (e.g. velocity with units of m/s), `DimensionalUnits` may be used to describe the system of units employed.

`FlowEquationSet` contains a description of the governing flow equations associated with the entire CGNS database. This structure contains information on the general class of governing equations (e.g. Euler or Navier-Stokes), equation sets required for closure, including turbulence modeling and equations of state, and constants associated with the equations.

`DataClass`, `DimensionalUnits`, `ReferenceState` and `FlowEquationSet` have special function in the CGNS hierarchy. They are globally applicable throughout the database, but their precedence may be superseded by local entities (e.g. within a given zone). The scope of these entities and the rules for determining precedence are treated in Section 6.3.

Globally relevant convergence history information is contained in `GlobalConvergenceHistory`. This convergence information includes total configuration forces, moments, and global residual and solution-change norms taken over all the zones.

Miscellaneous global data may be contained in the `IntegralData_t` list. Candidates for inclusion here are global forces and moments.

The `Family_t` data structure, defined in Section 12.6, is used to record geometry reference data. It may also include boundary conditions linked to geometry patches. For the purpose of defining material properties, families may also be defined for groups of elements. The family-mesh association is defined under the `Zone_t` and `BC_t` data structures by specifying the family name corresponding to a zone or a boundary patch.

6.2 Zone Structure Definition: `Zone_t`

The `Zone_t` structure contains all information pertinent to an individual zone. This information includes the zone type, the number of cells and vertices making up the grid in that zone, the physical coordinates of the grid vertices, grid motion information, the family, the flow solution, zone interface connectivity, boundary conditions, and zonal convergence history data. Zonal data may be recorded at multiple time steps or iterations. In addition, this structure contains a reference state, a set of flow equations and dimensional units that are all unique to the zone. For unstructured zones, the element connectivity may also be recorded.

```
ZoneType_t := Enumeration(
  Null,
  Structured,
  Unstructured,
  UserDefined ) ;

Zone_t< int CellDimension, int PhysicalDimension > :=
  {
  List( Descriptor_t Descriptor1 ... DescriptorN ) ;                (o)

  ZoneType_t ZoneType ;                                             (r)

  int[IndexDimension] VertexSize ;                                  (r)
  int[IndexDimension] CellSize ;                                    (r)
  int[IndexDimension] VertexSizeBoundary ;                          (o/d)

  List( GridCoordinates_t<IndexDimension, VertexSize>
        GridCoordinates MovedGrid1 ... MovedGridN ) ;               (o)

  List( Elements_t Elements1 ... ElementsN ) ;                      (o)

  List( RigidGridMotion_t RigidGridMotion1 ... RigidGridMotionN ) ; (o)

  List( ArbitraryGridMotion_t
        ArbitraryGridMotion1 ... ArbitraryGridMotionN ) ;          (o)

  FamilyName_t FamilyName ;                                         (o)

  List( FlowSolution_t<IndexDimension, VertexSize, CellSize>
        FlowSolution1 ... FlowSolutionN ) ;                        (o)

  List( DiscreteData_t<IndexDimension, VertexSize, CellSize>
        DiscreteData1 ... DiscreteDataN ) ;                        (o)

  List( IntegralData_t IntegralData1 ... IntegralDataN ) ;         (o)

  ZoneGridConnectivity_t<IndexDimension, CellDimension>
     ZoneGridConnectivity ;                                        (o)
```

```
ZoneBC_t<IndexDimension, PhysicalDimension> ZoneBC ;              (o)

ZoneIterativeData_t<NumberOfSteps> ZoneIterativeData ;            (o)

ReferenceState_t ReferenceState ;                                (o)

DataClass_t DataClass ;                                          (o)

DimensionalUnits_t DimensionalUnits ;                            (o)

FlowEquationSet_t<CellDimension> FlowEquationSet ;               (o)

ConvergenceHistory_t ZoneConvergenceHistory ;                    (o)

int Ordinal ;                                                    (o)
} ;
```

Notes

1. Default names for the `Descriptor_t`, `Elements_t`, `RigidGridMotion_t`, `ArbitraryGridMo-tion_t`, `FlowSolution_t`, `DiscreteData_t`, and `IntegralData_t` lists are as shown; users may choose other legitimate names. Legitimate names must be unique within a given instance of `Zone_t` and shall not include the names `DataClass`, `DimensionalUnits`, `Family-Name`, `FlowEquationSet`, `GridCoordinates`, `Ordinal`, `ReferenceState`, `ZoneBC`, `ZoneConvergenceHistory`, `ZoneGridConnectivity`, `ZoneIterativeData`, or `ZoneType`.
2. The original grid coordinates should have the name `GridCoordinates`. Default names for the remaining entities in the `GridCoordinates_t` list are as shown; users may choose other legitimate names, subject to the restrictions listed in the previous note.
3. `ZoneType`, `VertexSize`, and `CellSize` are the only required fields within the `Zone_t` structure.

`Zone_t` requires the parameters `CellDimension` and `PhysicalDimension`. `CellDimension`, along with the type of zone, determines `IndexDimension`; if the zone type is `Unstructured`, `IndexDimension = 1`, and if the zone type is `Structured`, `IndexDimension = CellDimension`. These three structure parameters identify the dimensionality of the grid-size arrays. One or more of them are passed on to the grid coordinates, flow solution, interface connectivity, boundary condition and flow-equation description structures.

`VertexSize` is the number of vertices in each index direction, and `CellSize` is the number of cells in each direction; for structured grids in 3-D, `CellSize = VertexSize - [1,1,1]`. `VertexSize` is the number of vertices defining 'the grid' or the domain (i.e. without rind points); `CellSize` is the number of cells on the interior of the domain. These two grid-size arrays are passed onto the grid-coordinate, flow-solution and discrete-data substructures.

If the nodes are sorted between internal nodes and boundary nodes, then the optional parameter `VertexSizeBoundary` must be set equal to the number of boundary nodes. If the nodes are sorted, the grid coordinate vector must first include the boundary nodes, followed by the internal nodes. By default, `VertexSizeBoundary` equals zero, meaning that the nodes are unsorted. This option is only useful for unstructured zones. For structured zones, `VertexSizeBoundary` always equals 0 in all index directions.

The `GridCoordinates_t` structure defines "the grid"; it contains the physical coordinates of the grid vertices, and may optionally contain physical coordinates of rind or ghost points. The original grid is contained in `GridCoordinates`. Additional `GridCoordinates_t` data structures are allowed, to store the grid at multiple time steps or iterations.

When the grid nodes are sorted, the `DataArray_t` in `GridCoordinates_t` lists first the data for the boundary nodes, then the data for the internal nodes.

The `Elements_t` data structure contains unstructured elements data such as connectivity, element type, parent elements, etc.

The `RigidGridMotion_t` and `ArbitraryGridMotion_t` data structures contain information defining rigid and arbitrary (i.e., deforming) grid motion.

`FamilyName` identifies to which family a zone belongs. Families may be used to define material properties.

Flow-solution quantities are contained in the list of `FlowSolution_t` structures. Each instance of the `FlowSolution_t` structure is only allowed to contain data at a single grid location (vertices, cell-centers, etc.); multiple `FlowSolution_t` structures are provided to store flow-solution data at different grid locations, to record different solutions at the same grid location, or to store solutions at multiple time steps or iterations. These structures may optionally contain solution data defined at rind points.

Miscellaneous discrete field data is contained in the list of `DiscreteData_t` structures. Candidate information includes residuals, fluxes and other related discrete data that is considered auxiliary to the flow solution. Likewise, miscellaneous zone-specific global data, other than reference-state data and convergence history information, is contained in the list of `IntegralData_t` structures. It is envisioned that these structures will be seldom used in practice but are provided nonetheless.

For unstructured zones only, the node-based `DataArray_t` vectors (`GridLocation = Vertex`) in `FlowSolution_t` or `DiscreteData_t` must follow exactly the same ordering as the `GridCoordinates` vector. If the nodes are sorted (`VertexSizeBoundary` $\neq 0$), the data on the boundary nodes must be listed first, followed by the data on the internal nodes. Note that the order in which the node-based data are recorded must follow exactly the node ordering in `GridCoordinates_t`, to be able to associate the data to the correct nodes. For element-based data (`GridLocation = CellCenter`), the `FlowSolution_t` or `DiscreteData_t` data arrays must list the data values for each element, in the same order as the elements are listed in `ElementConnectivity`.

All interface connectivity information, including identification of overset-grid holes, for a given zone is contained in `ZoneGridConnectivity`.

All boundary condition information pertaining to a zone is contained in `ZoneBC_t`.

The `ZoneIterativeData_t` data structure may be used to record pointers to zonal data at multiple time steps or iterations.

Reference-state data specific to an individual zone is contained in the `ReferenceState` structure.

`DataClass` defines the zonal default for the class of data contained in the zone and its substructures. If a zone contains dimensional data, `DimensionalUnits` may be used to describe the system of dimensional units employed.

If a set of flow equations are specific to a given zone, these may be described in `FlowEquationSet`. For example, if a single zone within the domain is inviscid, whereas all other are turbulent, then this zone-specific equation set could be used to describe the special zone.

`DataClass`, `DimensionalUnits`, `ReferenceState` and `FlowEquationSet` have special function in the hierarchy. They are applicable throughout a given zone, but their precedence may be superseded by local entities contained in the zone's substructures. If any of these entities are present within a given instance of `Zone_t`, they take precedence over the corresponding global entities contained in database's `CGNSBase_t` entity. These precedence rules are further discussed in Section 6.3.

Convergence history information applicable to the zone is contained in `ZoneConvergenceHistory`; this includes residual and solution-change norms.

`Ordinal` is user-defined and has no restrictions on the values that it can contain. It is included for backward compatibility to assist implementation of the CGNS system into applications whose I/O depends heavily on the numbering of zones. Since there are no restrictions on the values contained in `Ordinal` (or that `Ordinal` is even provided), there is no guarantee that the zones in an existing CGNS database will have sequential values from 1 to N without holes or repetitions. Use of `Ordinal` is discouraged and is on a user-beware basis.

6.3 Precedence Rules and Scope Within the Hierarchy

The dependence of a structure entity's information on data contained at higher levels of the hierarchy is typically explicitly expressed through structure parameters. For example, all arrays within `Zone_t` depend on the dimensionality of the computational grid. This dimensionality is passed down to a `Zone_t` entity through a structure parameter in the definition of `Zone_t`.

We have established an alternate dependency for a limited number of entities that is not explicitly stated in the structure type definitions. These special situations include entities for describing data class, system of dimensional units, reference states and flow equation sets. At each level of the hierarchy (where appropriate), entities for describing this information are defined, and if present they take precedence over all corresponding information existing at higher levels of the CGNS hierarchy. Essentially, we have established globally applicable data with provisions for recursively overriding it with local data.

Specifically, the entities that follow this alternate dependency are:

- `FlowEquationSet_t FlowEquationSet`,

- `ReferenceState_t ReferenceState`,

- `DataClass_t DataClass`,

- `DimensionalUnits_t DimensionalUnits`.

`FlowEquationSet` contains a description of the governing flow equations (see Section 10); `Refer-enceState` describes a set of reference state flow conditions (see Section 12.1); `DataClass` defines the class of data (e.g. dimensional or nondimensional—see Section 4.1 and Section 5); and `Dimen-sionalUnits` specifies the system of units used for dimensional data (see Section 4.3).

All of these entities may be defined within the highest level `CGNSBase_t` structure, and if present in a given database, establish globally applicable information; these may also be considered to be global defaults. Each of these four entities may also be defined within the `Zone_t` structure. If present in a given instance of `Zone_t`, they supersede the global data and establish new defaults which apply only within that zone. For example, if a different set of flow equations is solved within a given zone than is solved in the rest of the flowfield, then this can be conveyed through `FlowEquationSet`.

In this case, one `FlowEquationSet` entity would be placed within `CGNSBase_t` to state the globally applicable flow equations, and a second `FlowEquationSet` entity would be placed within the odd zone (within its instance of `Zone_t`); this second `FlowEquationSet` entity supersedes the first only within the odd zone.

In addition to its presence in `CGNSBase_t` and `Zone_t`, `ReferenceState` may also be defined within the boundary-condition structure types to establish reference states applicable to one or more boundary-condition patches. Actually, `ReferenceState` entities can be defined at several levels of the boundary-condition hierarchy to allow flexibility in setting the appropriate reference state conditions (see Section 9.1 and subsequent sections).

`DataClass` and `DimensionalUnits` are used within entities describing data arrays (see the `DataArray_t` type definition in Section 5.1). They classify the data and specify its system of units if dimensional. If these entities are absent from a particular instance of `DataArray_t`, the information is derived from appropriate global data. `DataClass` and `DimensionalUnits` are also declared in all intermediate structure types that directly or indirectly contain `DataArray_t` entities. Examples include `GridCoordinates_t` (Section 7.1), `FlowSolution_t` (Section 7.5), `BC_t` (Section 9.3) and `ReferenceState_t` (Section 12.1). The same precedence rules apply—lower-level entities supersede higher-level entities.

It is envisioned that in practice, the use of globally applicable data will be the norm rather than the exception. It provides a measure of economy throughout the CGNS database in many situations. For example, when creating a database where the vast majority of data arrays are dimensional and use a consistent set of units, `DataClass` and `DimensionalUnits` can be set appropriately at the `CGNSBase_t` level and thereafter omitted when outputting data.

7 Grid Coordinates, Elements, and Flow Solutions

This section defines structure types for describing the grid coordinates, element data, and flow solution data pertaining to a zone. Entities of each of the structure types defined in this section are contained in the Zone_t structure (see Section 6.2).

7.1 Grid Coordinates Structure Definition: GridCoordinates_t

The physical coordinates of the grid vertices are described by the GridCoordinates_t structure. This structure contains a list for the data arrays of the individual components of the position vector. It also provides a mechanism for identifying rind-point data included within the position-vector arrays.

```
GridCoordinates_t< int IndexDimension, int VertexSize[IndexDimension] > :=
  {
  List( Descriptor_t Descriptor1 ... DescriptorN ) ;                       (o)

  Rind_t<IndexDimension> Rind ;                                            (o/d)

  List( DataArray_t<DataType, IndexDimension, DataSize[]>
        DataArray1 ... DataArrayN ) ;                                      (o)

  DataClass_t DataClass ;                                                  (o)

  DimensionalUnits_t DimensionalUnits ;                                    (o)
  } ;
```

Notes

1. Default names for the Descriptor_t and DataArray_t lists are as shown; users may choose other legitimate names. Legitimate names must be unique within a given instance of Grid-Coordinates_t and shall not include the names DataClass, DimensionalUnits, or Rind.
2. There are no required fields for GridCoordinates_t. Rind has a default if absent; the default is equivalent to having a Rind structure whose RindPlanes array contains all zeros (see Section 4.8).
3. The structure parameter DataType must be consistent with the data stored in the DataArray_t substructures (see Section 5.1).
4. For unstructured zones, rind planes are not meaningful and should not be used.

GridCoordinates_t requires two structure parameters: IndexDimension identifies the dimensionality of the grid-size arrays, and VertexSize is the number of vertices in each index direction excluding rind points. For unstructured zones, IndexDimension is always 1 and VertexSize is the total number of vertices.

The grid-coordinates data is stored in the list of DataArray_t entities; each DataArray_t structure entity may contain a single component of the position vector (e.g. three separate DataArray_t entities are used for x, y, and z). Standardized data-name identifiers for the grid coordinates are described in Annex A.

Rind is an optional field that indicates the number of rind planes included in the grid-coordinates data for structured zones. If Rind is absent, then the DataArray_t structure entities contain only 'core' vertices of a zone; 'core' refers to all interior and bounding vertices of a zone – VertexSize is the number of 'core' vertices. 'Core' vertices in a zone are assumed to begin at [1,1,1] (for a structured zone in 3-D) and end at VertexSize. If Rind is present, it will provide information on the number of 'rind' points in addition to the 'core' points that are contained in the DataArray_t structures.

DataClass defines the default class for data contained in the DataArray_t entities. For dimensional grid coordinates, DimensionalUnits may be used to describe the system of units employed. If present, these two entities take precedence over the corresponding entities at higher levels of the CGNS hierarchy. The rules for determining precedence of entities of this type are discussed in Section 6.3. An example that uses these grid-coordinate defaults is shown in Section 7.2.

FUNCTION DataSize[]:

return value: one-dimensional int array of length IndexDimension
dependencies: IndexDimension, VertexSize[], Rind

GridCoordinates_t requires a single structure function, named DataSize, to identify the array sizes of the grid-coordinates data. A function is required for the following reasons:

- the entire grid, including both 'core' and 'rind' points, is stored in the DataArray_t entities;
- the DataArray_t structure is simple in that it doesn't know anything about 'core' versus 'rind' data; it just knows that it contains data of some given size;
- to make all the DataArray_t entities syntactically consistent in their size (i.e. by syntax entities containing x, y and z must have the same dimensionality and dimension sizes), the size of the array is passed onto the DataArray_t structure as a parameter.

```
if (Rind is absent) then
  {
  DataSize[] = VertexSize[] ;
  }
else if (Rind is present) then
  {
  DataSize[] = VertexSize[] + [a + b,...] ;
  }
```

where RindPlanes = [a,b,...] (see Section 4.8 for the definition of RindPlanes).

7.2 Grid Coordinates Examples

This section contains examples of grid coordinates. These examples show the storage of the grid-coordinate data arrays, as well as different mechanisms for describing the class of data and the system of units or normalization.

Example 7-A: Cartesian Coordinates for a 2-D Structured Grid

Cartesian coordinates for a 2-D grid of size 17×33; the data arrays include only core vertices, and the coordinates are in units of feet.

```
!  IndexDimension = 2
!  VertexSize = [17,33]
GridCoordinates_t<2, [17,33]> GridCoordinates =
  {{
  DataArray_t<real, 2, [17,33]> CoordinateX =
    {{
    Data(real, 2, [17,33]) = ((x(i,j), i=1,17), j=1,33) ;

    DataClass_t DataClass = Dimensional ;

    DimensionalUnits_t DimensionalUnits =
      {{
      MassUnits        = Null ;
      LengthUnits      = Foot ;
      TimeUnits        = Null ;
      TemperatureUnits = Null ;
      AngleUnits       = Null ;
      }} ;
    }} ;

  DataArray_t<real, 2, [17,33]> CoordinateY =
    {{
    Data(real, 2, [17,33]) = ((y(i,j), i=1,17), j=1,33) ;

    DataClass_t DataClass = Dimensional ;

    DimensionalUnits_t DimensionalUnits =
      {{
      MassUnits        = Null ;
      LengthUnits      = Foot ;
      TimeUnits        = Null ;
      TemperatureUnits = Null ;
      AngleUnits       = Null ;
      }} ;
    }} ;
  }} ;
```

From Annex A, the identifiers for x and y are CoordinateX and CoordinateY, respectively, and both have a data type of real. The value of DataClass in CoordinateX and CoordinateY indicate the data is dimensional, and DimensionalUnits specifies the appropriate units are feet. The

DimensionalExponents entity is absent from both CoordinateX and CoordinateY; the information that x and y are lengths can be inferred from the data-name identifier conventions in Annex A.1.

Note that FORTRAN multidimensional array indexing is used to store the data; this is reflected in the FORTRAN-like implied do-loops for x(i,j) and y(i,j).

Since the dimensional units for both x and y are the same, an alternate approach is to set the data class and system of units using DataClass and DimensionalUnits at the GridCoordinates_t level, and eliminate this information from each instance of DataArray_t.

```
GridCoordinates_t<2, [17,33]> GridCoordinates =
  {{
  DataClass_t DataClass = Dimensional ;

  DimensionalUnits_t DimensionalUnits =
    {{
    MassUnits          = Null ;
    LengthUnits        = Foot ;
    TimeUnits          = Null ;
    TemperatureUnits   = Null ;
    AngleUnits         = Null ;
    }} ;

  DataArray_t<real, 2, [17,33]> CoordinateX =
    {{
    Data(real, 2, [17,33]) = ((x(i,j), i=1,17), j=1,33) ;
    }} ;

  DataArray_t<real, 2, [17,33]> CoordinateY =
    {{
    Data(real, 2, [17,33]) = ((y(i,j), i=1,17), j=1,33) ;
    }} ;
  }} ;
```

Since the DataClass and DimensionalUnits entities are not present in CoordinateX and CoordinateY, the rules established in Section 5.2.1 dictate that DataClass and DimensionalUnits specified at the GridCoordinates_t level be used to retrieve the information.

Example 7-B: Cylindrical Coordinates for a 3-D Structured Grid

Cylindrical coordinates for a 3-D grid whose core size is $17 \times 33 \times 9$. The grid contains a single plane of rind on the minimum and maximum k faces. The coordinates are nondimensional.

```
!  IndexDimension = 3
!  VertexSize = [17,33,9]
GridCoordinates_t<3, [17,33,9]> GridCoordinates =
  {{
```

```
Rind_t<3> Rind =
  {{
  int[6] RindPlanes = [0,0,0,0,1,1] ;
  }} ;


! DataType = real
! IndexDimension = 3
! DataSize = VertexSize + [0,0,2] = [17,33,11]
DataArray_t<real, 3, [17,33,11]> CoordinateRadius =
  {{
  Data(real, 3, [17,33,11]) = (((r(i,j,k), i=1,17), j=1,33), k=0,10) ;

  DataClass_t DataClass = NormalizedByUnknownDimensional ;
  }} ;


DataArray_t<real, 3, [17,33,11]> CoordinateZ     = {{ }} ;
DataArray_t<real, 3, [17,33,11]> CoordinateTheta = {{ }} ;
}} ;
```

The value of RindPlanes specifies two rind planes on the minimum and maximum k faces. These rind planes are reflected in the structure function DataSize which is equal to the number of core vertices plus two in the k dimension. The value of DataSize is passed to the DataArray_t entities. The value of DataClass indicates the data is nondimensional. Note that if DataClass is set as NormalizedByUnknownDimensional at a higher level (CGNSBase_t or Zone_t), then it is not needed here.

Note that the entities CoordinateZ and CoordinateTheta are abbreviated.

Example 7-C: Cartesian Coordinates for a 3-D Unstructured Grid

Cartesian grid coordinates for a 3-D unstructured zone where VertexSize is 15.

```
GridCoordinates_t<1, 15> GridCoordinates =
  {{

  ! DataType = real
  ! IndexDimension = 1
  ! DataSize = VertexSize = 15
  DataArray_t<real, 1, 15> CoordinateX =
    {{
    Data(real, 1, 15) = (x(i), i=1,15) ;
    }} ;


  DataArray_t<real, 1, 15> CoordinateY =
    {{
    Data(real, 1, 15) = (y(i), i=1,15) ;
    }} ;
```

```
DataArray_t<real, 1, 15> CoordinateZ =
  {{
  Data(real, 1, 15) = (z(i), i=1,15) ;
  }} ;
}} ;
```

7.3   Elements Structure Definition: Elements_t

The Elements_t data structure is required for unstructured zones, and contains the element connectivity data, the element type, the element range, the parent elements data, and the number of boundary elements.

```
Elements_t :=
  {
  List( Descriptor_t Descriptor1 ... DescriptorN ) ;                       (o)

  IndexRange_t ElementRange ;                                              (r)

  int ElementSizeBoundary ;                                                (o/d)

  ElementType_t ElementType ;                                              (r)

  DataArray_t<int, 1, ElementDataSize> ElementConnectivity ;               (r)

  DataArray_t<int, 2, [ElementSize, 4]> ParentData;                        (o)
  } ;
```

Notes

1. Default names for the Descriptor_t list are as shown; users may choose other legitimate names. Legitimate names must be unique within a given instance of Elements_t and shall not include the names ElementConnectivity, ElementRange, or ParentData.
2. IndexRange_t, ElementType_t, and ElementConnectivity_t are the required fields within the Elements_t structure.

ElementRange contains the index of the first and last elements defined in ElementConnectivity. The elements are indexed with a global numbering system, starting at 1, for all element sections under a given Zone_t data structure. They are also listed as a continuous list of element numbers within any single element section. Therefore the number of elements in a section is:

```
ElementSize = ElementRange.end - ElementRange.start + 1
```

The element indices are used for the boundary condition and zone connectivity definition.

ElementSizeBoundary indicates if the elements are sorted, and how many boundary elements are recorded. By default, ElementSizeBoundary is set to zero, indicating that the elements are not sorted. If the elements are sorted, ElementSizeBoundary is set to the number of elements at the boundary. Consequently:

```
ElementSizeInterior = ElementSize - ElementSizeBoundary
```

ElementType_t is an enumeration of the supported element types:

```
ElementType_t := Enumeration(
   Null, NODE, BAR_2, BAR_3,
   TRI_3, TRI_6, QUAD_4, QUAD_8, QUAD_9,
   TETRA_4, TETRA_10, PYRA_5, PYRA_14,
   PENTA_6, PENTA_15, PENTA_18,
   HEXA_8, HEXA_20, HEXA_27, MIXED, NGON_n, UserDefined );
```

Section 3.3 illustrates the convention for element numbering.

For all element types except type MIXED, ElementConnectivity contains the list of nodes for each element. If the elements are sorted, then it must first list the connectivity of the boundary elements, then that of the interior elements.

$$
\begin{aligned}
\texttt{ElementConnectivity} = &\texttt{Node1}_1, \texttt{Node2}_1, \ldots \texttt{NodeN}_1,\\
&\texttt{Node1}_2, \texttt{Node2}_2, \ldots \texttt{NodeN}_2,\\
&\ldots\\
&\texttt{Node1}_M, \texttt{Node2}_M, \ldots \texttt{NodeN}_M
\end{aligned}
$$

When the section ElementType is MIXED, the data array ElementConnectivity contains one extra integer per element, to hold each individual element type:

$$
\begin{aligned}
\texttt{ElementConnectivity} = &\texttt{Etype}_1, \texttt{Node1}_1, \texttt{Node2}_1, \ldots \texttt{NodeN}_1,\\
&\texttt{Etype}_2, \texttt{Node1}_2, \texttt{Node2}_2, \ldots \texttt{NodeN}_2,\\
&\ldots\\
&\texttt{Etype}_M, \texttt{Node1}_M, \texttt{Node2}_M, \ldots \texttt{NodeN}_M
\end{aligned}
$$

ElementDataSize indicates the size (number of integers) of the array ElementConnectivity. For all element types except type MIXED, the ElementDataSize is given by:

```
ElementDataSize = ElementSize * NPE[ElementType]
```

In the case of MIXED element section, ElementDataSize is given by:

$$
\texttt{ElementDataSize} = \sum_{n=start}^{end} \left(\texttt{NPE[ElementType}_n\texttt{]} + 1 \right)
$$

NPE[ElementType] is a function returning the number of nodes for the given ElementType. For example, NPE[HEXA_8]=8.

For face elements in 3D, or bar element in 2D, four more data may be saved for each element — the corresponding parents' element number, and the face position within these parent elements. At the boundaries, the second parent is set to zero.

NGON_n is used to express a polygon of n nodes. In order to record the number of nodes of any ngons, the ElementType must be set to NGON_n + Nnodes. For example, for an element type NGON_n composed of 25 nodes, one would set the ElementType to NGON_n + 25.

7.4 Elements Examples

This section contains two examples of elements definition in CGNS. In both cases, the unstructured zone contains 15 tetrahedral and 10 hexahedral elements.

Example 7-D: Unstructured Elements, Separate Element Types

In this first example, the elements are written in two separate sections, one for the tetrahedral elements and one for the hexahedral elements.

```
Zone_t UnstructuredZone =
  {{
  Elements_t TetraElements =
    {{
    IndexRange_t ElementRange = [1,15] ;

    int ElementSizeBoundary = 10 ;

    ElementType_t ElementType = TETRA_4 ;

    DataArray_t<int, 1, NPE[TETRA_4]×15> ElementConnectivity =
      {{
      Data(int, 1, NPE[TETRA_4]×15) = (node(i,j), i=1,NPE[TETRA_4], j=1,15) ;
      }} ;

    DataArray_t<int, 1, 4×15> ParentData =
      {{
      Data(int, 1, 4×15) = (parent1(j),parent2(j),face1(j),face2(j), j=1,15) ;
      }} ;
    }} ;
  Elements_t HexaElements =
    {{
    IndexRange_t ElementRange = [16,25] ;

    int ElementSizeBoundary = 0 ;
```

```
ElementType_t ElementType = HEXA_8 ;

DataArray_t<int, 1, NPE[HEXA_8]×10> ElementConnectivity =
  {{
  Data(int, 1, NPE[HEXA_8]×10) = (node(i,j), i=1,NPE[HEXA_8], j=1,10) ;
  }} ;

DataArray_t<int, 1, 4×10> ParentData =
  {{
  Data(int, 1, 4×10) = (parent1(j),parent2(j),face1(j),face2(j), j=1,10) ;
  }} ;
  }} ;
}} ;
```

For elements 1 through 10, `parent2` and `face2` are set to zero since these are boundary elements.

Example 7-E: Unstructured Elements, Element Type `MIXED`

In this second example, the same unstructured zone described in Example 7-D is written in a single element section of type `MIXED` (i.e., an unstructured grid composed of mixed elements).

```
Zone_t UnstructuredZone =
  {{
  Elements_t MixedElementsSection =
    {{
    IndexRange_t ElementRange = [1,25] ;

    ElementType_t ElementType = MIXED ;

    DataArray_t<int, 1, ElementDataSize> ElementConnectivity =
      {{
      Data(int, 1, ElementDataSize) = (etype(j),(node(i,j),
          i=1,NPE[etype(j)]), j=1,25) ;
      }} ;

    DataArray_t<int, 1, 4×25> ParentData =
      {{
      Data(int, 1, 4×25) = (parent1(j),parent2(j),face1(j),face2(j), j=1,25) ;
      }} ;
    }} ;
  }} ;
```

7.5 Flow Solution Structure Definition: `FlowSolution_t`

The flow solution within a given zone is described by the `FlowSolution_t` structure. This structure contains a list for the data arrays of the individual flow-solution variables, as well as identifying the grid location of the solution. It also provides a mechanism for identifying rind-point data included within the data arrays.

```
FlowSolution_t< int IndexDimension, int VertexSize[IndexDimension],
              int CellSize[IndexDimension] > :=
  {
  List( Descriptor_t Descriptor1 ... DescriptorN ) ;                    (o)

  GridLocation_t GridLocation ;                                         (o/d)

  Rind_t<IndexDimension> Rind ;                                         (o/d)

  List( DataArray_t<DataType, IndexDimension, DataSize[]>
      DataArray1 ... DataArrayN ) ;                                     (o)

  DataClass_t DataClass ;                                               (o)

  DimensionalUnits_t DimensionalUnits ;                                 (o)
  } ;
```

Notes

1. Default names for the `Descriptor_t` and `DataArray_t` lists are as shown; users may choose other legitimate names. Legitimate names must be unique within a given instance of `FlowSolution_t` and shall not include the names `DataClass`, `DimensionalUnits`, `GridLocation`, or `Rind`.
2. There are no required fields for `FlowSolution_t`. `GridLocation` has a default of `Vertex` if absent. `Rind` also has a default if absent; the default is equivalent to having an instance of `Rind` whose `RindPlanes` array contains all zeros (see Section 4.8).
3. The structure parameter `DataType` must be consistent with the data stored in the `DataArray_t` structure entities (see Section 5.1); `DataType` is `real` for all flow-solution identifiers defined in Annex A.
4. For unstructured zones: rind planes are not meaningful and should not be used; `GridLocation` options are limited to `Vertex` or `CellCenter`, meaning that solution data may only be expressed at these locations; and the data arrays must follow the node ordering if `GridLocation = Vertex`, and the element ordering if `GridLocation = CellCenter`.

`FlowSolution_t` requires three structure parameters; `IndexDimension` identifies the dimensionality of the grid-size arrays, and `VertexSize` and `CellSize` are the number of 'core' vertices and cells, respectively, in each index direction. For unstructured zones, `IndexDimension` is always 1.

The flow solution data is stored in the list of `DataArray_t` entities; each `DataArray_t` structure entity may contain a single component of the solution vector. Standardized data-name identifiers for the flow-solution quantities are described in Annex A. The field `GridLocation` specifies the location of the solution data with respect to the grid; if absent, the data is assumed to coincide with grid vertices (i.e. GridLocation = Vertex). All data within a given instance of FlowSolution_t must reside at the same grid location.

Rind is an optional field for structured zones that indicates the number of rind planes included in the data. Its purpose and function are identical to those described in Section 7.1. Note, however, that the Rind in this structure is independent of the Rind contained in GridCoordinates_t. They are not required to contain the same number of rind planes. Also, the location of any flow-solution rind points is assumed to be consistent with the location of the 'core' flow solution points (e.g. if GridLocation = CellCenter, rind points are assumed to be located at fictitious cell centers).

DataClass defines the default class for data contained in the DataArray_t entities. For dimensional flow solution data, DimensionalUnits may be used to describe the system of units employed. If present these two entities take precedence over the corresponding entities at higher levels of the CGNS hierarchy. The rules for determining precedence of entities of this type are discussed in Section 6.3.

FUNCTION DataSize[]:

return value: one-dimensional int array of length IndexDimension
dependencies: IndexDimension, VertexSize[], CellSize[], GridLocation, Rind

The function DataSize[] is the size of flow solution data arrays. If Rind is absent then DataSize represents only the 'core' points; it will be the same as VertexSize or CellSize depending on GridLocation. The definition of the function DataSize[] is as follows:

```
if (Rind is absent) then
  {
  if (GridLocation = Vertex) or (GridLocation is absent)
    {
    DataSize[] = VertexSize[] ;
    }
  else if (GridLocation = CellCenter) then
    {
    DataSize[] = CellSize[] ;
    }
  }
else if (Rind is present) then
  {
  if (GridLocation = Vertex) or (GridLocation is absent) then
    {
    DataSize[] = VertexSize[] + [a + b,...] ;
    }
  else if (GridLocation = CellCenter)
    {
    DataSize[] = CellSize[] + [a + b,...] ;
    }
  }
```

where RindPlanes = [a,b,...] (see Section 4.8 for the definition of RindPlanes).

7.6 Flow Solution Example

This section contains an example of the flow solution entity, including the designation of grid location and rind planes and data-normalization mechanisms.

Example 7-F: Flow Solution

Conservation-equation variables (ρ, ρu, ρv and ρe_0) for a 2-D grid of size 11×5. The flowfield is cell-centered with two planes of rind data. The density, momentum and stagnation energy (ρe_0) data is nondimensionalized with respect to a freestream reference state whose quantities are dimensional. The freestream density and pressure are used for normalization; these values are 1.226 kg/m^3 and 1.0132×10^5 N/m^2 (standard atmosphere conditions). The data-name identifier conventions for the conservation-equation variables are Density, MomentumX, MomentumY and EnergyStagnation-Density.

```
!   IndexDimension = 2
!   VertexSize = [11,5]
!   CellSize = [10,4]
FlowSolution_t<2, [11,5], [10,4]> FlowExample =
  {{
  GridLocation_t GridLocation = CellCenter ;

  Rind_t<2> Rind =
    {{
    int[4] RindPlanes = [2,2,2,2] ;
    }} ;

  DataClass_t DataClass = NormalizedByDimensional ;

  DimensionalUnits_t DimensionalUnits =
    {{
    MassUnits        = Kilogram ;
    LengthUnits      = Meter ;
    TimeUnits        = Second ;
    TemperatureUnits = Null ;
    AngleUnits       = Null ;
    }} ;

!   DataType = real
!   Dimension = 2
!   DataSize = CellSize + [4,4] = [14,8]
DataArray_t<real, 2, [14,8]> Density =
    {{
    Data(real, 2, [14,8]) = ((rho(i,j), i=-1,12), j=-1,6) ;

    DataConversion_t DataConversion =
```

```
    {{
    ConversionScale  = 1.226 ;
    ConversionOffset = 0 ;
    }} ;

  DimensionalExponents_t DimensionalExponents =
    {{
    MassExponent        = +1 ;
    LengthExponent      = -3 ;
    TimeExponent        =  0 ;
    TemperatureExponent =  0 ;
    AngleExponent       =  0 ;
    }} ;
  }} ;

DataArray_t<real, 2, [14,8]> MomentumX =
  {{
  Data(real, 2, [14,8]) = ((rho_u(i,j), i=-1,12), j=-1,6) ;

  DataConversion_t DataConversion =
    {{
    ConversionScale  = 352.446 ;
    ConversionOffset = 0 ;
    }} ;
  }} ;

DataArray_t<real, 2, [14,8]> MomentumY =
  {{
  Data(real, 2, [14,8]) = ((rho_v(i,j), i=-1,12), j=-1,6) ;

  DataConversion_t DataConversion =
    {{
    ConversionScale  = 352.446 ;
    ConversionOffset = 0 ;
    }} ;
  }} ;

DataArray_t<real, 2, [14,8]> EnergyStagnationDensity =
  {{
  Data(real, 2, [14,8]) = ((rho_e0(i,j), i=-1,12), j=-1,6) ;

  DataConversion_t DataConversion =
    {{
    ConversionScale  = 1.0132e+05 ;
    ConversionOffset = 0 ;
```

```
      }} ;
    }} ;
  }} ;
```

The value of `GridLocation` indicates the data is at cell centers, and the value of `RindPlanes` specifies two rind planes on each face of the zone. The resulting value of the structure function `DataSize` is the number of cells plus four in each coordinate direction; this value is passed to each of the `DataArray_t` entities.

Since the data are all nondimensional and normalized by dimensional reference quantities, this information is stated in `DataClass` and `DimensionalUnits` at the `FlowSolution_t` level rather than attaching the appropriate `DataClass` and `DimensionalUnits` to each `DataArray_t` entity. It could possibly be at even higher levels in the heirarchy. The contents of `DataConversion` are in each case the denominator of the normalization; this is ρ_∞ for density, $\sqrt{p_\infty \rho_\infty}$ for momentum, and p_∞ for stagnation energy. The dimensional exponents are specified for density. For all the other data, the dimensional exponents are to be inferred from the data-name identifiers.

Note that no information is provided to identify the actual reference state or indicate that it is freestream. This information is not needed for data manipulations involving renormalization or changing the units of the converted raw data.

8   Multizone Interface Connectivity

This section defines structures for describing multizone interface connectivity for 1-to-1 abutting, mismatched abutting, and overset type interfaces. The different types of zone interfaces are described in Section 3.4. All interface connectivity information pertaining to a given zone is grouped together in a `ZoneGridConnectivity_t` structure entity; this in turn is contained in a zone structure entity (see the definition of `Zone_t` in Section 6.2).

Before presentation of the structure definitions, a few design features require comment. All indices used to describe interfaces are the dimensionality (`IndexDimension`) of the grid, even when they are used to describe lower-dimensional zonal boundaries for abutting interfaces. The alternative for structured zones that was not chosen is to use lower-dimensional indices for lower-dimensional interfaces (e.g. for a 3-D grid, use two-dimensional indices for describing grid planes that are interfaces). Both alternatives offer trade-offs. The lower-dimensional indices require cyclic notation conventions and additional identification of face location; whereas, full-dimensional indices result in one redundant index component when describing points along a grid plane. We decided that full-dimensional indices would be more usable and less error prone in actual implementation.

A major consequence of this decision is that connectivity information for describing mismatched abutting interfaces and overset interfaces can be merged into a single structure, `GridConnectivity_t` (see Section 8.4 below). In fact, this single structure type can be used to describe all zonal interfaces.

A second design choice was to duplicate all 1-to-1 abutting interface information within the CGNS database. It is possible to describe a given 1-to-1 interface with a single set of connectivity data. In contrast, mismatched and overset interfaces require different connectivity information when the roles of receiver and donor zones are interchanged. Therefore, a given mismatched or overset interface requires two sets of connectivity data within the database. The decision to force two sets of connectivity data (one contained in each of the `Zone_t` entities for the two adjacent zones) for each 1-to-1 interface makes the connectivity structures for all interface types look and function similarly. It also fits better with the zone-by-zone hierarchy chosen for the CGNS database. The minor penalty in data duplication was deemed worth the advantages gained.

8.1   Zonal Connectivity Structure Definition: `ZoneGridConnectivity_t`

All multizone interface grid connectivity information pertaining to a given zone is contained in the `ZoneGridConnectivity_t` structure. This includes abutting interfaces (1-to-1 and general mismatched), overset-grid interfaces, and overset-grid holes.

```
ZoneGridConnectivity_t< int IndexDimension, int CellDimension > :=
  {
  List( Descriptor_t Descriptor1 ... DescriptorN ) ;                    (o)

  List( GridConnectivity1to1_t<IndexDimension>
        GridConnectivity1to11 ... GridConnectivity1to1N ) ;             (o)
```

```
List( GridConnectivity_t<IndexDimension, CellDimension>
      GridConnectivity1 ... GridConnectivityN ) ;                    (o)

List( OversetHoles_t<IndexDimension>
      OversetHoles1 ... OversetHolesN ) ;                            (o)
} ;
```

Notes

1. Default names for the `Descriptor_t`, `GridConnectivity1to1_t`, `GridConnectivity_t` and `OversetHoles_t` lists are as shown; users may choose other legitimate names. Legitimate names must be unique within a given instance of `ZoneGridConnectivity_t`.
2. All lists within the `ZoneGridConnectivity_t` structure may be empty.

`ZoneGridConnectivity_t` requires two structure parameters, `IndexDimension`, which is passed onto all connectivity substructures, and `CellDimension`, which is passed to `GridConnectivity_t` only.

Connectivity information for 1-to-1 or matched multizone interfaces is contained in the `GridConnectivity1to1_t` structure. Abutting and overset connectivity is contained in the `GridConnectivity_t` structure, and overset-grid holes are identified in the `OversetHoles_t` structure.

8.2 1-to-1 Interface Connectivity Structure Definition: `GridConnectivity1to1_t`

`GridConnectivity1to1_t` only applies to structured zones interfacing with structured donors and whose interface is a logically rectangular region. It contains connectivity information for a multizone interface patch that is abutting with 1-to-1 matching between adjacent zone indices (also referred to as C0 connectivity). An interface patch is the subrange of the face of a zone that touches one and only one other zone. This structure identifies the subrange of indices for the two adjacent zones that make up the interface and gives an index transformation from one zone to the other. It also identifies the name of the adjacent zone.

All the interface patches for a given zone are contained in the `ZoneGridConnectivity_t` entity for that zone. If a face of a zone touches several other zones (say N), then N different instances of the `GridConnectivity1to1_t` structure must be included in the zone to describe each separate interface patch. This convention requires that a single interface patch be described twice in the database—once for each adjacent zone.

```
GridConnectivity1to1_t< int IndexDimension > :=
  {
  List( Descriptor_t Descriptor1 ... DescriptorN ) ;                 (o)

  int[IndexDimension] Transform ;                                    (o/d)

  IndexRange_t<IndexDimension> PointRange ;                          (r)
  IndexRange_t<IndexDimension> PointRangeDonor ;                     (r)
```

```
Identifier(Zone_t) ZoneDonorName ;                              (r)

int Ordinal ;                                                   (o)
} ;
```

Notes

1. Default names for the `Descriptor_t` list are as shown; users may choose other legitimate names. Legitimate names must be unique within a given instance of `GridConnectivity-1to1_t` and shall not include the names `PointRange`, `PointRangeDonor`, `Transform`, or `Ordinal`.
2. If `Transform` is absent, then its default value is [+1,+2,+3].
3. `ZoneDonorName` must be equated to a zone identifier within the current CGNS database (i.e. it must be equal to one of the `Zone_t` identifiers contained in the current `CGNSBase_t` entity).
4. Beginning indices of `PointRange` and `PointRangeDonor` must coincide (i.e. must be the same physical point); ending indices of `PointRange` and `PointRangeDonor` must also coincide.
5. Elements of `Transform` must be signed integers in the range $-$IndexDimension, ..., +IndexDimension; element magnitudes may not be repeated. In 3-D allowed elements are 0, ± 1, ± 2, ± 3.

`PointRange` contains the subrange of indices that makes up the interface patch in the current zone (i.e. that `Zone_t` entity that contains the given instance of `GridConnectivity1to1_t`). `PointRangeDonor` contains the interface patch subrange of indices for the adjacent zone (whose identifier is given by `ZoneDonorName`). By convention the indices contained in `PointRange` and `PointRangeDonor` refer to vertices.

`Transform` contains a short-hand notation for the transformation matrix describing the relation between indices of the two adjacent zones. The transformation matrix itself has rank `IndexDimension` and contains elements $+1$, -1 and 0; it is orthonormal and its inverse is its transpose. The transformation matrix (T) works as follows: if `Index1` and `Index2` are the indices of a given point on the interface, where `Index1` is in the current zone and `Index2` is in the adjacent zone, then their relationship is,

```
Index2 = T.(Index1 - Begin1) + Begin2

Index1 = Transpose[T].(Index2 - Begin2) + Begin1
```

where the '.' notation indicates matrix-vector multiply. `Begin1` and `End1` are the subrange indices contained in `PointRange`, and `Begin2` and `End2` are the subrange indices contained in `PointRange-Donor`.

The short-hand notation used in `Transform` is as follows: each element shows the image in the adjacent zone's face of a positive index increment in the current zone's face. The first element is the image of a positive increment in i; the second element is the image of an increment in j;

and the third (in 3-D) is the image of an increment in k on the current zone's face. For 3-D, the transformation matrix T is constructed from $\mathtt{Transform} = [\pm a, \pm b, \pm c]$ as follows:

$$
\mathtt{T} = \begin{bmatrix}
\text{sgn}(a)\,\text{del}(a-1) & \text{sgn}(b)\,\text{del}(b-1) & \text{sgn}(c)\,\text{del}(c-1) \\
\text{sgn}(a)\,\text{del}(a-2) & \text{sgn}(b)\,\text{del}(b-2) & \text{sgn}(c)\,\text{del}(c-2) \\
\text{sgn}(a)\,\text{del}(a-3) & \text{sgn}(b)\,\text{del}(b-3) & \text{sgn}(c)\,\text{del}(c-3)
\end{bmatrix},
$$

where,

$$
\text{sgn}(x) \equiv \begin{cases} +1, & \text{if } x \geq 0 \\ -1, & \text{if } x < 0 \end{cases}
\qquad
\text{del}(x-y) \equiv \begin{cases} 1, & \text{if abs}(x) = \text{abs}(y) \\ 0, & \text{otherwise} \end{cases}
$$

For example, $\mathtt{Transform} = [-2, +3, +1]$ gives the transformation matrix,

$$
\mathtt{T} = \begin{bmatrix} 0 & 0 & +1 \\ -1 & 0 & 0 \\ 0 & +1 & 0 \end{bmatrix}.
$$

For establishing relationships between adjacent and current zone indices lying on the interface itself, one of the elements of $\mathtt{Transform}$ is superfluous since one component of both interface indices remains constant. It is therefore acceptable to set that element of $\mathtt{Transform}$ to zero.

If the transformation matrix is used for continuation of computational coordinates into the adjacent zone (e.g. to find the location of a rind point in the adjacent zone), then all elements of $\mathtt{Transform}$ are needed. If the above mentioned superfluous element is set to zero, it can be easily regenerated from $\mathtt{PointRange}$ and $\mathtt{PointRangeDonor}$ and the grid sizes of the two zones. This is done by determining the faces represented by $\mathtt{PointRange}$ and $\mathtt{PointRangeDonor}$ (i.e. i-min, i-max, j-min, etc.). If one is a minimum face and the other a maximum face, then the sign of the missing element in $\mathtt{Transform}$ is '+', and the value of the missing element in the transformation matrix (T) is $+1$. If the faces are both minimums or are both maximums, the sign is '−'. Next, the position and magnitude of the element in $\mathtt{Transform}$, and hence the row and column in the transformation matrix, is given by the combinations of i, j and k faces for the two. For example, if $\mathtt{PointRange}$ represents a j-min or j-max face and $\mathtt{PointRangeDonor}$ represents an i-min or i-max face, then the missing element's position in $\mathtt{Transform}$ is 2 and its magnitude is 1 (i.e. $\mathtt{Transform} = [*, \pm 1, *]$).

Note also that the transform matrix and the two index pairs overspecify the interface patch. For example, $\mathtt{End2}$ can be obtained from $\mathtt{Transform}$, $\mathtt{Begin1}$, $\mathtt{End1}$ and $\mathtt{Begin2}$.

$\mathtt{Ordinal}$ is user-defined and has no restrictions on the values that it can contain. It is included for backward compatibility to assist implementation of the CGNS system into applications whose I/O depends heavily on the numbering of zone interfaces. Since there are no restrictions on the values contained in $\mathtt{Ordinal}$ (or that $\mathtt{Ordinal}$ is even provided), there is no guarantee that the interfaces in an existing CGNS database will have sequential values from 1 to N without holes or repetitions. Use of $\mathtt{Ordinal}$ is discouraged and is on a user-beware basis.

8.3   1-to-1 Interface Connectivity Examples

This section contains two examples of structure entities for describing the connectivity for structured-zone 1-to-1 abutting multizone interfaces. Annex B contains additional examples of 1-to-1 interfaces.

Example 8-A: 1-to-1 Abutting of Complete Faces

Two zones have the same orientation; zone 1 is $9 \times 17 \times 11$ and zone 2 is $9 \times 17 \times 21$. The k-max face of zone 1 abuts the k-min face of zone 2. Contained in the structure entities of zone 1 is the following interface structure:

```
GridConnectivity1to1_t<3> Zone1/ZoneGridConnectivity/KMax =
  {{
  int[3] Transform = [1,2,3] ;
  IndexRange_t<3> PointRange =
    {{
    int[3] Begin = [1,1,11] ;
    int[3] End   = [9,17,11] ;
    }} ;
  IndexRange_t<3> PointRangeDonor =
    {{
    int[3] Begin = [1,1,1] ;
    int[3] End   = [9,17,1] ;
    }} ;
  Identifier(Zone_t) ZoneDonorName = Zone2 ;
  }} ;
```

Contained in the structure entities of zone 2 is the following:

```
GridConnectivity1to1_t<3> Zone2/ZoneGridConnectivity/KMin =
  {{
  int[3] Transform = [1,2,3] ;
  IndexRange_t<3> PointRange =
    {{
    int[3] Begin = [1,1,1] ;
    int[3] End   = [9,17,1] ;
    }} ;
  IndexRange_t<3> PointRangeDonor =
    {{
    int[3] Begin = [1,1,11] ;
    int[3] End   = [9,17,11] ;
    }} ;
  Identifier(Zone_t) ZoneDonorName = Zone1 ;
  }} ;
```

This example assumes zones 1 and 2 have the identifiers **Zone1** and **Zone2**, respectively.

Example 8-B: 1-to-1 Abutting, Complete Face to a Subset of a Face

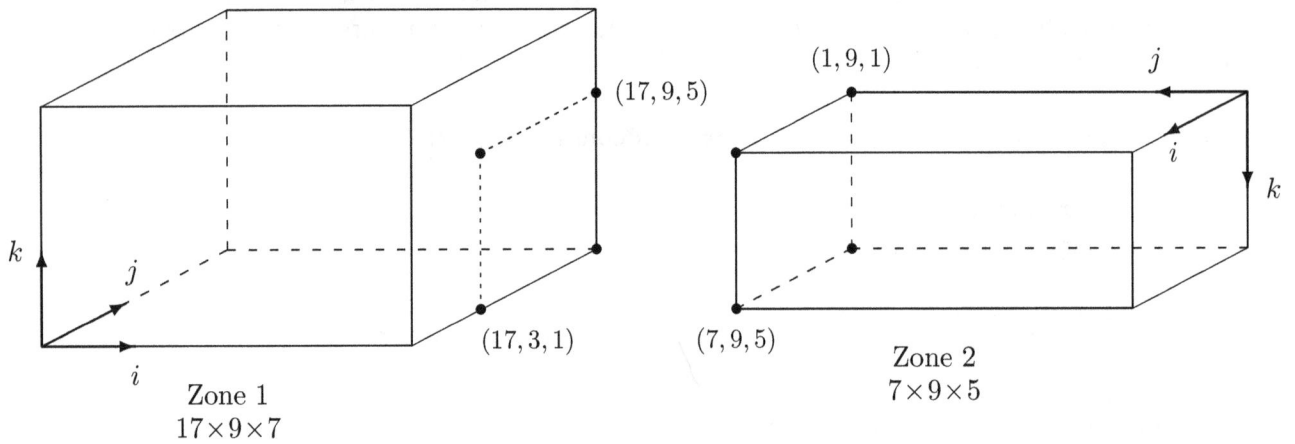

Figure 3: Example Interface for 1-to-1 Connectivity

Figure 3 shows a more complex 1-to-1 abutting interface, where the entire j-max face of zone 2 coincides with a subset of the i-max face of zone 1. This situation would result in the following connectivity structures:

```
GridConnectivity1to1_t<3> Zone1/ZoneGridConnectivity/IMax =
  {{
  int[3] Transform = [-2,-1,-3] ;
  IndexRange_t<3> PointRange =
    {{
    int[3] Begin = [17,3,1] ;
    int[3] End   = [17,9,5] ;
    }} ;
  IndexRange_t<3> PointRangeDonor =
    {{
    int[3] Begin = [7,9,5] ;
    int[3] End   = [1,9,1] ;
    }} ;
  Identifier(Zone_t) ZoneDonorName = Zone2 ;
  }} ;

GridConnectivity1to1_t<3> Zone2/ZoneGridConnectivity/JMax =
  {{
  int[3] Transform = [-2,-1,-3] ;
  IndexRange_t<3> PointRange =
    {{
    int[3] Begin = [1,9,1] ;
    int[3] End   = [7,9,5] ;
    }} ;
```

```
IndexRange_t<3> PointRangeDonor =
  {{
  int[3] Begin = [17,9,5] ;
  int[3] End   = [17,3,1] ;
  }} ;
Identifier(Zone_t) ZoneDonorName = Zone1 ;
}} ;
```

This example also assumes zones 1 and 2 have the identifiers **Zone1** and **Zone2**, respectively. Note that the index transformation matrix for both this and the previous examples is symmetric; hence, the value of **Transform** is identical for both members of the interface pair. In general this will not always be the case.

8.4 General Interface Connectivity Structure Definition: GridConnectivity_t

GridConnectivity_t contains connectivity information for generalized multizone interfaces, and may be used for any mix of structured and unstructured zones. Its purpose is to describe mismatched-abutting and overset interfaces, but can also be used for 1-to-1 abutting interfaces.

For abutting interfaces that are not 1-to-1, also referred to as patched or mismatched, an interface patch is the subrange of the face of a zone that touches one and only one other zone. This structure identifies the subrange of indices (or array of indices) that make up the interface and gives their image in the adjacent (donor) zone. It also identifies the name of the adjacent zone. If a given face of a zone touches several (say N) adjacent zones, then N different instances of GridConnectivity_t are needed to describe all the interfaces. For a single abutting interface, two instances of GridConnectivity_t are needed in the database – one for each adjacent zone.

For overset interfaces, this structure identifies the fringe points of a given zone that lie in one and only one other zone. If the fringe points of a zone lie in several (say N) overlapping zones, then N different instances of GridConnectivity_t are needed to describe the overlaps. It is possible with overset grids that a single fringe point may actually lie in several overlapping zones (though in typical usage, linkage to only one of the overlapping zones is kept). There is no restriction against a given fringe point being contained within multiple instances of GridConnectivity_t; therefore, this structure allows the description of a single fringe point lying in several overlapping zones.

```
GridConnectivityType_t := Enumeration(
  Null,
  Overset,
  Abutting,
  Abutting1to1,
  UserDefined ) ;


GridConnectivity_t< int IndexDimension, int CellDimension > :=
  {
  List( Descriptor_t Descriptor1 ... DescriptorN ) ;                    (o)
```

```
      GridConnectivityType_t GridConnectivityType ;                    (o/d)

      GridLocation_t GridLocation ;                                    (o/d)

      IndexRange_t<IndexDimension> PointRange ;                        (o:r)
      IndexArray_t<IndexDimension, PointListSize, int>  PointList ;    (r:o)
      IndexArray_t<IndexDimension, PointListSize, int>  PointListDonor ;  (o:r)
      IndexArray_t<IndexDimension, PointListSize, int>  CellListDonor ;   (r:o)

      Identifier(Zone_t) ZoneDonorName ;                               (r)

      DataArray_t <real, 2, [CellDimension, PointListSize]> InterpolantsDonor (r:o)

      int Ordinal ;                                                    (o)
      } ;
```

Notes

1. Default names for the `Descriptor_t` list are as shown; users may choose other legitimate names. Legitimate names must be unique within a given instance of `GridConnectivity_t` and shall not include the names `CellListDonor`, `GridConnectivityType`, `GridLocation`, `InterpolantsDonor`, `Ordinal`, `PointList`, `PointListDonor`, or `PointRange`.
2. `ZoneDonorName` must be equated to a zone identifier within the current CGNS database (i.e. it must be equal to one of the `Zone_t` identifiers contained in the current `CGNSBase_t` entity).
3. If `GridConnectivityType` is absent, then its default value is `Overset`.
4. If `GridLocation` is absent, then its default value is `Vertex`.
5. One of `PointRange` and `PointList` must be specified, but not both.
6. If `PointRange` is specified, then an index ordering convention is needed to map receiver-zone grid points to donor-zone grid points. FORTRAN multidimensional array ordering is used.
7. If `GridConnectivityType` is `Abutting1to1` or `Abutting`, then `PointRange` or `PointList` must define points associated with a face subrange (if the zone is structured, all points must be in a single computational grid plane); the donor-zone grid locations defined by `PointListDonor` or `CellListDonor` must also be associated with a face subrange.
8. Either `PointListDonor` alone, or `CellListDonor` plus `InterpolantsDonor`, must be used. The use of `PointListDonor` is restricted to `Abutting1to1`, whereas `CellListDonor` plus `InterpolantsDonor` can be used for any interface type.

The type of multizone interface connectivity may be `Overset`, `Abutting`, or `Abutting1to1`. Overset refers to zones that overlap; for a 3-D configuration the overlap is a 3-D region. `Abutting` refers to zones that abut or touch, but do not overlap (other than the vertices and faces that make up the interface). `Abutting1to1` is a special case of abutting interfaces where grid lines are continuous across the interface and all vertices on the interface are shared by the two adjacent zones. See Section 3.4 for a description of the three different types of interfaces.

The interface grid points within the receiver zone may be specified by `PointRange` if they constitute a logically rectangular region (e.g. an abutting interface where an entire face of the receiver zone abuts with a part of a face of the donor zone). In all other cases, `PointList` should be used to list the receiver-zone grid points making up the interface. For a structured-to-structured interface, all indices in `PointRange` or `PointList` should have one index element in common (see note 7).

`GridLocation` identifies the location of indices within the receiver zone described by `PointRange` or `PointList`; it also identifies the location of indices defined by `PointListDonor` in the donor zone. This allows the flexibility to specify grid locations other than vertices. `GridLocation` does *not* apply to `CellListDonor` or `InterpolantsDonor`. The `CellListDonor` is always an index or indices that define a particular cell or element, while the `InterpolantsDonor` defines an interpolation value relative to the cell/element *vertices*.

`PointListDonor` may only be used when the interface is `Abutting1to1`. It contains the images of all the receiver-zone interface points in the donor zone. If the zone is structured, all indices in `PointListDonor` should have one index element in common.

For mismatched or overset interfaces, the zone connectivity is defined using the combination of `CellListDonor` and `InterpolantsDonor`. `CellListDonor` contains the list of donor cells in which each node of the receiver zone can be located. `InterpolantsDonor` contains the interpolation factors to locate the receiver nodes in the donor cells. `InterpolantsDonor` may be thought of as bi- or tri-linear interpolants (depending on `CellDimension`) in the cell of the donor zone.

`Ordinal` is user-defined and has no restrictions on the values that it can contain. It is included for backward compatibility to assist implementation of the CGNS system into applications whose I/O depends heavily on the numbering of zone interfaces. Since there are no restrictions on the values contained in `Ordinal` (or that `Ordinal` is even provided), there is no guarantee that the interfaces for a given zone in an existing CGNS database will have sequential values from 1 to N without holes or repetitions. Use of `Ordinal` is discouraged and is on a user-beware basis.

FUNCTION `PointListSize`:

return value: `int`
dependencies: `PointRange, PointList`

`PointListDonor`, `CellListDonor`, and `InterpolantsDonor` require the function `PointListSize`, to identify the length of the array. If `PointRange` is specified by `GridConnectivity_t`, then `PointListSize` is obtained from the number of grid points (inclusive) between the beginning and ending indices of `PointRange`. If `PointList` is specified by by `GridConnectivity_t`, then `PointListSize` is actually a user input during creation of the database; it is the length of the array `PointList` whose elements are also user inputs (by 'user' we mean the application code that is generating the CGNS database).

By definition, the `PointList` and `PointListDonor` arrays have the same size, and this size should be stored along with the arrays in their respective `IndexArray_t` structures (this is done in the ADF implementation). `PointListSize` was chosen to be a structure function, rather than a separate element of `GridConnectivity_t` for the following reasons: first, it is redundant if `PointRange` is specified; and second, it leads to redundant storage if `PointList` is specified, since the value of `PointListSize` is also stored within the `PointList` structure.

This situation has somewhat of a precedent within the SIDS definitions. The structure `Descrip-`

tor_t contains a string of unspecified length. Yet in actual implementation, the (string) length is a function of the descriptor string itself and should be stored along with the string.

8.5 Overset Grid Holes Structure Definition: OversetHoles_t

Grid connectivity for overset grids must also include 'holes' within zones, where any solution states are ignored or 'turned off', because they are solved for in some other overlapping zone. The structure OversetHoles_t specifies those points within a given zone that make up a hole (or holes), and applies to both structured and unstructured zones. Grid points specified in this structure are equivalent to those with IBLANK=0 in the PLOT3D format.

```
OversetHoles_t< int IndexDimension > :=
  {
  List( Descriptor_t Descriptor1 ... DescriptorN ) ;                    (o)

  GridLocation_t GridLocation ;                                         (o/d)

  List( IndexRange_t<IndexDimension>
    PointRange, PointRange2 ... PointRangeN ) ;                         (o)

  IndexArray_t<IndexDimension, PointListSize, int> PointList ;          (o)
  } ;
```

Notes

1. Default names for the Descriptor_t and IndexRange_t lists are as shown; users may choose other legitimate names. Legitimate names must be unique within a given instance of Overset-Holes_t and shall not include the names GridLocation or PointList.
2. If GridLocation is absent, then its default value is Vertex.

The location of grid indices specified in PointList and the PointRange list is given by GridLocation.

The grid points making up a hole within a zone may be specified by PointRange if they constitute a logically rectangular region. If the hole points constitute a (small) set of possibly overlapping logically rectangular regions, then they may be specified by the list PointRange, PointRange2, etc. The more general alternate is to use PointList to list all grid points making up the hole(s) within a zone. Using the list of PointRange specifications, or using PointRange in combination with PointList, may result in a given hole being specified more than once.

FUNCTION PointListSize:

return value: int
dependencies: PointList

`OversetHoles_t` requires one structure function, `PointListSize`, to identify the length of the `PointList` array. `PointListSize` is a user input (see discussion on function `PointListSize` in previous section).

9   Boundary Conditions

This section is an attempt to unify boundary-condition specifications within Navier-Stokes codes. The structures and conventions developed are a compromise between simplicity and generality. It is imperative that they be easy to use initially, but that they are general enough to provide future flexibility and extensibility.

This section may be somewhat daunting initially. It is suggested that the reader refer to the several, well-explained examples presented in Section 9.9 during study of the following sections to help resolve any questions and confusions that might arise.

The difficulty with boundary conditions is that there is such a wide variety used, and even a single boundary-condition equation is often implemented differently in different codes. Some boundary conditions, such as a symmetry plane, are fairly well defined. Other boundary conditions are much looser in their definition and implementation. An inflow boundary is a good example. It is generally accepted how many solution quantities should be specified at an inflow boundary (from mathematical well-posedness arguments), but what those quantities are will change with the class of flow problems (e.g. internal flows vs. external flows), and they will also change from code to code.

An additional difficulty for CFD analysis is that in some situations different boundary-condition equations are applied depending on local flow conditions. Any boundary where the flow can change from inflow to outflow or supersonic to subsonic is a candidate for flow-dependent boundary-condition equations.

These difficulties have molded the design of our boundary-condition specification structures and conventions. We define boundary-condition types (Section 9.6) that establish the equations to be enforced. However, for those more loosely defined boundary conditions, such as inflow/outflow, the boundary-condition type merely establishes general guidelines on the equations to be imposed. Augmenting (and superseding) the information provided by the boundary-condition type is precisely defined boundary-condition solution data. We rely on our conventions for data-name identifiers to identify the exact quantities involved in the boundary conditions; these data-name identifier conventions are presented in Annex A.

One flexibility that is provided by this approach is that boundary-condition information can easily be built during the course of an analysis. For example, during grid-generation phases minimal information (e.g. the boundary-condition type) may be given. Then prior to running of the flow solver, more specific boundary-condition information, such as Dirichlet or Neumann data, may be added to the database.

An additional flexibility provided by the structures of this section is that both uniform and non-uniform boundary-condition data can be described within the same framework.

We realize that most current codes allow little or no flexibility in choosing solution quantities to specify for a given boundary-condition type. We also realize the coding effort involved with checking for consistency between I/O specifications and internal boundary-condition routines. To make these boundary-condition structures more palatable initially, we adopt the convention that if no solution quantities are specified for a given boundary-condition type, then the code is free to enforce any appropriate boundary condition (see Section 9.8).

Currently, there are no boundary-condition structures defined for abutting or overset interfaces, unless they involve cases of symmetry or degeneracy. There is also no separate boundary-condition structure for periodic boundary conditions (i.e. when a zone interfaces with itself).

In the sections to follow, the definitions of boundary-condition structures are first presented in Section 9.1 through Section 9.5. Boundary-condition types are then discussed in detail in Section 9.6, including a description of the boundary-condition equations to be enforced for each type; this section also describes the distinction between boundary-condition types that impose a set of equations regardless of local flow conditions and those that impose different sets of boundary-condition equations depending on the local flow solution. The rules for matching boundary-condition types and the appropriate sets of boundary-condition equations are next discussed in Section 9.7. Details of specifying data to be imposed in boundary-condition equations are provided in Section 9.8. Finally, Section 9.9 presents several examples of boundary conditions.

9.1 Boundary Condition Structures Overview

Prior to presenting the detailed boundary condition structures, we give a brief overview of the hierarchy used to describe boundary conditions.

Boundary conditions are classified as either fixed or flow-dependent. Fixed boundary conditions enforce a given set of boundary-condition equations regardless of flow conditions; whereas, flow-dependent boundary conditions enforce different sets of boundary-condition equations depending on local flow conditions. We incorporate both fixed and flow-dependent boundary conditions into a uniform framework. This allows all boundary conditions to be described in a similar manner. We consider this functionally superior than separately treating fixed and flow-dependent boundary conditions, even though the latter allows a simpler description mechanism for fixed boundary conditions. The current organization also makes sense considering the fact that flow-dependent boundary conditions are composed of multiple sets of fixed boundary conditions.

Figure 4 depicts the hierarchy used for prescribing a single boundary condition. Each boundary condition includes a type that describes the general equations to enforce, a patch specification, and a collection of data sets. The minimum required information for any boundary condition is the patch specification and the boundary-condition type (indicated by "BC type (compound)" in the figure). This minimum information is similar to that used in many existing flow solvers.

Generality in prescribing equations to enforce and their associated boundary-condition data is provided in the optional data sets. Each data set contains all boundary condition data required for a given fixed or simple boundary condition. Each data set is also tagged with a boundary-condition type. For fixed boundary conditions, the hierarchical tree contains a single data set, and the two boundary-condition types shown in Figure 4 are identical. Flow-dependent or compound boundary conditions contain multiple data sets, each to be applied separately depending on local flow conditions. The compound boundary-condition type describes the general flow-dependent boundary conditions, and each data set contains associated simple boundary-condition types. For example, a farfield boundary condition would contain four data sets, where each applies to the different combinations of subsonic and supersonic inflow and outflow. Boundary-condition types are described in Section 9.6 and Section 9.7.

Within a single data set, boundary condition data is grouped by equation type into Dirichlet and Neumann data. The lower leaves of Figure 4 show data for generic flow-solution quantities α and

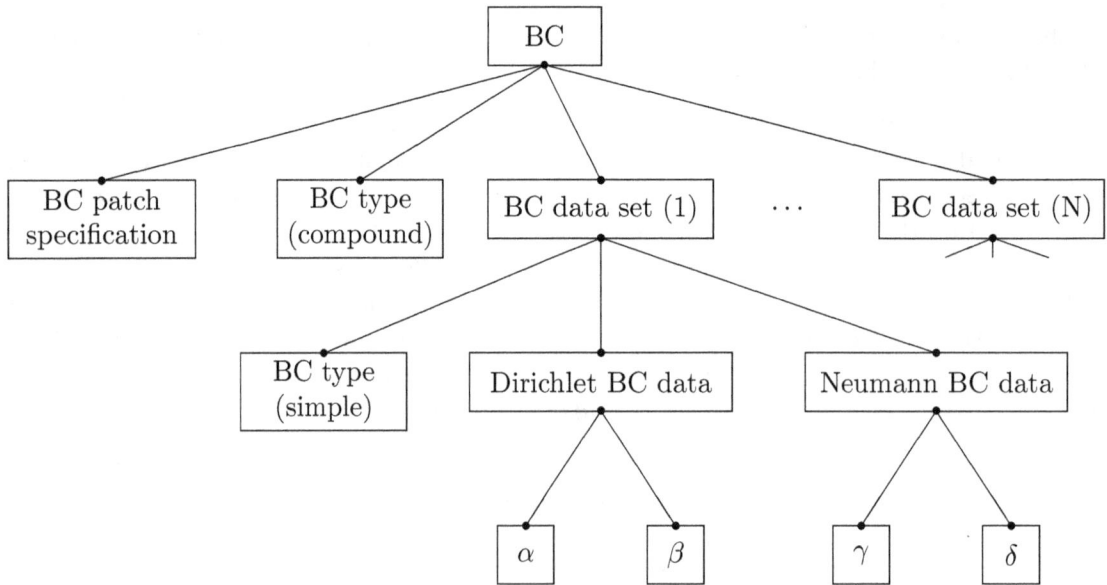

Figure 4: Hierarchy for Boundary Condition Structures

β to be applied in Dirichlet conditions, and data for γ and δ to be applied in Neumann boundary conditions. `DataArray_t` entities are employed to store these data and to identify the specific flow variables they are associated with.

In situations where the data sets (or any information contained therein) are absent from a given boundary-condition hierarchy, flow solvers are free to impose any appropriate boundary conditions. Although not pictured in Figure 4, it is also possible to specify the reference state from which the flow solver should extract the boundary-condition data.

9.2 Zonal Boundary Condition Structure Definition: ZoneBC_t

All boundary-condition information pertaining to a given zone is contained in the `ZoneBC_t` structure.

```
ZoneBC_t< int IndexDimension, int PhysicalDimension > :=
  {
  List( Descriptor_t Descriptor1 ... DescriptorN ) ;              (o)

  List( BC_t<IndexDimension, int PhysicalDimension> BC1 ... BCN ) ;   (o)

  ReferenceState_t ReferenceState ;                               (o)

  DataClass_t DataClass ;                                         (o)

  DimensionalUnits_t DimensionalUnits ;                           (o)
  } ;
```

Notes

1. Default names for the `Descriptor_t` and `BC_t` lists are as shown; users may choose other legitimate names. Legitimate names must be unique within a given instance of `ZoneBC_t` and shall not include the names `DataClass`, `DimensionalUnits`, or `ReferenceState`.
2. All lists within a `ZoneBC_t` structure entity may be empty.

`ZoneBC_t` requires two structure parameters, `IndexDimension` and `PhysicalDimension`, which are passed onto all `BC_t` substructures.

Boundary-condition information for a single patch is contained in the `BC_t` structure. All boundary-condition information pertaining to a given zone is contained in the list of `BC_t` structure entities. If a zone contains N boundary-condition patches, then N separate instances of `BC_t` must be provided in the `ZoneBC_t` entity for the zone.

Reference data applicable to all boundary conditions of a zone is contained in the `ReferenceState` structure. `DataClass` defines the zonal default for the class of data contained in the boundary conditions of a zone. If the boundary conditions contain dimensional data, `DimensionalUnits` may be used to describe the system of dimensional units employed. If present, these three entities take precedence of all corresponding entities at higher levels of the hierarchy. These precedence rules are further discussed in Section 6.3.

Reference-state data is useful for situations where boundary-condition data is not provided, and flow solvers are free to enforce any appropriate boundary condition equations. The presense of `ReferenceState` at this level or below specifies the appropriate flow conditions from which the flow solver should extract its boundary-condition data. For example, a engine nozzle exit boundary condition usually imposes a stagnation pressure (or some other stagnation quantity) different from freestream. The nozzle-exit stagnation quantities could be specified in an instance of `ReferenceState` at this level or below in lieu of providing explicit Dirichlet or Neumann data (see Section 9.8).

9.3 Boundary-Condition Structure Definition: `BC_t`

`BC_t` contains boundary-condition information for a single BC surface patch of a zone. A BC patch is the subrange of the face of a zone where a given boundary condition is applied.

The structure contains a boundary-condition type, as well as one or more sets of boundary-condition data that are used to define the boundary-condition equations to be enforced on the BC patch. For most boundary conditions, a single data set is all that is needed. The structure also contains information describing the normal vector to the BC surface patch.

```
BC_t< int IndexDimension, int PhysicalDimension > :=
  {
  List( Descriptor_t  Descriptor1 ... DescriptorN ) ;                    (o)

  BCType_t  BCType ;                                                      (r)
```

```
GridLocation_t GridLocation ;                                          (o/d)

IndexRange_t<IndexDimension>  PointRange ;                             (o:r)
IndexArray_t<IndexDimension, ListLength, int>  PointList ;            (r:o)

int[IndexDimension]  InwardNormalIndex ;                              (o)

IndexArray_t<PhysicalDimension, ListLength, real> InwardNormalList ;  (o)

List( BCDataSet_t<ListLength> BCDataSet1 ... BCDataSetN ) ;           (o)

FamilyName_t FamilyName ;                                             (o)

ReferenceState_t ReferenceState ;                                    (o)

DataClass_t DataClass ;                                              (o)

DimensionalUnits_t DimensionalUnits ;                               (o)

int Ordinal ;                                                        (o)
} ;
```

Notes

1. Default names for the `Descriptor_t` and `BCDataSet_t` lists are as shown; users may choose other legitimate names. Legitimate names must be unique within a given instance of `BC_t` and shall not include the names `DataClass`, `DimensionalUnits`, `FamilyName`, `GridLocation`, `InwardNormalIndex`, `InwardNormalList`, `Ordinal`, `PointList`, `PointRange` or `ReferenceState`.

2. `GridLocation` may be set to `Vertex`, `IFaceCenter`, `JFaceCenter`, `KFaceCenter`, or `FaceCenter`, indicating that `PointList` or `PointRange` refer to vertices or cell faces. If `GridLocation` is absent, then its default value is `Vertex`.
   When `GridLocation` is set to `Vertex`, then `PointList` or `PointRange` refer to node indices, for both structured and unstructured grids. When `GridLocation` is set to `FaceCenter`, then `PointList` or `PointRange` refer to face elements. Face elements are indexed using different methods depending if the zone is structured or unstructured. For a structured zone, face elements are indexed using the minimum of the connecting vertex indices, as described in Section 3.2. For an unstructured zone, face elements are indexed using their element numbering, as defined in the `Elements_t` data structures.

3. One of `PointRange` and `PointList` must be specified, but not both. They must describe a face subrange. `PointRange` and `PointList` refer to either vertices or cell faces, depending on whether `GridLocation` is set to `Vertex` or `FaceCenter`.

4. `InwardNormalIndex` is only an option for structured grids. For unstructured grid boundaries, it should not be used. `InwardNormalIndex` may have only one nonzero element, whose sign indicates the computational-coordinate direction of the BC patch normal; this normal points

into the interior of the zone.

5. `InwardNormalList` contains a list of vectors normal to the BC patch pointing into the interior of the zone. It is a function of `PhysicalDimension` and `ListLength`. The vectors are located at the vertices of the BC patch when `GridLocation` is set to `Vertex`. For face center data (`GridLocation = FaceCenter`), the vectors are located at the cell-face midpoints. The vectors are not required to have unit magnitude.

6. If `PointRange` and `InwardNormalList` are specified, then an ordering convention is needed for indices on the BC patch. An ordering convention is also needed if `PointRange` is specified and local data is present in the `BCDataSet_t` substructures. FORTRAN multidimensional array ordering is used.

`BCType` specifies the boundary-condition type, which gives general information on the boundary-condition equations to be enforced. `BCType_t` is defined in Section 9.6 along with the meanings of all the `BCType` values.

The BC patch grid points may be specified by `PointRange` if they constitute a logically rectangular region. In all other cases, `PointList` should be used to list the vertices or cell faces making up the BC patch.

Some boundary conditions require a normal direction to be specified in order to be properly imposed. For structured zones a computational-coordinate normal can be derived from `PointRange` or `PointList` by examining redundant index components. Alternatively, for structured zones this information can be provided directly by `InwardNormalIndex`. From Note 4, this vector points into the zone and can have only one non-zero element. For exterior faces of a zone in 3-D, `InwardNormalIndex` should take the following values:

Face	InwardNormalIndex	Face	InwardNormalIndex
i-min	$[+1, 0, 0]$	i-max	$[-1, 0, 0]$
j-min	$[0, +1, 0]$	j-max	$[0, -1, 0]$
k-min	$[0, 0, +1]$	k-max	$[0, 0, -1]$

The physical-space normal vectors of the BC patch may be described by `InwardNormalList`; these are located at vertices or cell faces, consistent with `PointRange`, `PointList`, and `GridLocation`. `InwardNormalList` is listed as an optional field because it is not always needed to enforce boundary conditions, and the physical-space normals of a BC patch can usually be constructed from the grid. However, there are some situations, such as grid-coordinate singularity lines, where `InwardNormalList` becomes a required field, because it cannot be generated from other information.

The `BC_t` structure provides for a list of boundary-condition data sets, described in the next section. In general, the proper `BCDataSet_t` instance to impose on the BC patch is determined by the `BCType` association table (Table 4 on p. 87). The mechanics of determining the proper data set to impose is described in Section 9.7.

For a few boundary conditions, such as a symmetry plane or polar singularity, the value of `BCType` completely describes the equations to impose, and no instances of `BCDataSet_t` are needed. For 'simple' boundary conditions, where a single set of Dirichlet and/or Neumann data is applied,

AIAA R-101-2002

a single BCDataSet_t will likely appear (although this is not a requirement). For 'compound' boundary conditions, where the equations to impose are dependent on local flow conditions, several instances of BCDataSet_t will likely appear; the procedure for choosing the proper data set is more complex as described in Section 9.7.

FamilyName identifies the family to which the boundary belongs. Family names link the mesh boundaries to the CAD surfaces. (See Section 12.6.) Boundary conditions may also be defined directly on families. In this case, the BCType must be FamilySpecified. If, under a BC_t structure, both FamilyName_t and BCType_t are present, and the BCType is *not* FamilySpecified, then the BCType which *is* specified takes precedence over any BCType which might be stored in a FamilyBC_t structure under the specified Family_t.

Reference data applicable to the boundary conditions of a BC patch is contained in the ReferenceState structure. DataClass defines the default for the class of data contained in the boundary conditions. If the boundary conditions contain dimensional data, DimensionalUnits may be used to describe the system of dimensional units employed. If present, these three entities take precedence of all corresponding entities at higher levels of the hierarchy. These precedence rules are further discussed in Section 6.3.

Ordinal is user-defined and has no restrictions on the values that it can contain. It is included for backward compatibility to assist implementation of the CGNS system into applications whose I/O depends heavily on the numbering of BC patches. Since there are no restrictions on the values contained in Ordinal (or that Ordinal is even provided), there is no guarantee that the BC patches for a given zone in an existing CGNS database will have sequential values from 1 to N without holes or repetitions. Use of Ordinal is discouraged and is on a user-beware basis.

FUNCTION ListLength:

return value: int
dependencies: PointRange, PointList

BC_t requires the structure function ListLength, which is the number of vertices or cell faces making up the BC patch. If PointRange is specified, then ListLength is obtained from the number of points (inclusive) between the beginning and ending indices of PointRange. If PointList is specified, then ListLength is the number of indices in the list of points. In this situation, ListLength becomes a user input along with the indices of the list PointList. By 'user' we mean the application code that is generating the CGNS database.

ListLength is also the number of elements in the list InwardNormalList and is passed into the BCDataSet_t substructures, where it is used to determine the length of BC data arrays. Note that syntactically PointList and InwardNormalList must have the same number of elements.

9.4   Boundary Condition Data Set Structure Definition: BCDataSet_t

BCDataSet_t contains Dirichlet and Neumann data for a single set of boundary-condition equations. Its intended use is for simple boundary-condition types, where the equations imposed do not depend on local flow conditions.

```
BCDataSet_t< int ListLength > :=
  {
  List( Descriptor_t  Descriptor1 ... DescriptorN ) ;                (o)

  BCTypeSimple_t  BCTypeSimple ;                                     (r)

  BCData_t<ListLength>  DirichletData ;                              (o)
  BCData_t<ListLength>  NeumannData ;                                (o)

  ReferenceState_t ReferenceState ;                                  (o)

  DataClass_t DataClass ;                                            (o)

  DimensionalUnits_t DimensionalUnits ;                              (o)
  } ;
```

Notes

1. Default names for the `Descriptor_t` list are as shown; users may choose other legitimate names. Legitimate names must be unique within a given instance of `BCDataSet_t` and shall not include the names `DataClass`, `DimensionalUnits`, `DirichletData`, `NeumannData` or `ReferenceState`.
2. `BCTypeSimple` is the only required field. All other fields are optional and the `Descriptor_t` list may be empty.

`BCDataSet_t` requires the structure parameter `ListLength`, which is used to control the length of data arrays in the `Dirichlet` and `Neumann` substructures for data that is defined at vertices or face centers of the BC patch.

`BCTypeSimple` specifies the boundary-condition type, which gives general information on the boundary-condition equations to be enforced. `BCTypeSimple_t` is defined in Section 9.6 along with the meanings of all the `BCTypeSimple` values. `BCTypeSimple` is also used for matching boundary condition data sets as discussed in Section 9.7.

`GridLocation` is defined under `BC_t`, and specifies the location of local data arrays (if any) provided in `DirichletData` and `NeumannData`.

Boundary-condition data is separated by equation type into Dirichlet and Neumann conditions. Dirichlet boundary conditions impose the value of the given variables, whereas Neumann boundary conditions impose the normal derivative of the given variables. The mechanics of specifying Dirichlet and Neumann data for boundary conditions is covered in Section 9.8.

The substructures `DirichletData` and `NeumannData` contain boundary-condition data which may be constant over the BC patch or defined locally at each vertex or cell face of the patch. By design of `BCDataSet_t`, local boundary-condition data may be defined either at vertices or boundary face centers; this is governed by the value of `GridLocation` under `BC_t`. Rather than pass this location information into the `DirichletData` and `NeumannData` substructures, the total length of any local data arrays is instead passed using the `ListLength` structure function.

Reference quantities applicable to the set of boundary-condition data are contained in the **Reference-State** structure. **DataClass** defines the default for the class of data contained in the boundary-condition data. If the boundary conditions contain dimensional data, **DimensionalUnits** may be used to describe the system of dimensional units employed. If present, these three entities take precedence of all corresponding entities at higher levels of the hierarchy. These precedence rules are further discussed in Section 6.3.

9.5 Boundary Condition Data Structure Definition: BCData_t

BCData_t contains a list of variables and associated data for boundary-condition specification. Each variable may be given as global data (i.e. a scalar) or local data defined at each grid point of the BC patch. By convention all data specified in a given instance of **BCData_t** is to be used in the same *type* of boundary-condition equation. For example, the use of separate **BCData_t** substructures for Dirichlet and Neumann equations in the **BCDataSet_t** structure of the previous section.

```
BCData_t< int ListLength > :=
  {
  List( Descriptor_t  Descriptor1 ... DescriptorN ) ;              (o)

  List( DataArray_t<DataType, 1, 1>
        DataGlobal1 ... DataGlobalN ) ;                            (o)

  List( DataArray_t<DataType, 1, ListLength>
        DataLocal1 ... DataLocalN ) ;                             (o)

  DataClass_t DataClass ;                                         (o)

  DimensionalUnits_t DimensionalUnits ;                           (o)
  } ;
```

Notes

1. Default names for the **Descriptor_t** and **DataArray_t** lists are as shown; users may choose other legitimate names. Legitimate names must be unique within a given instance of **BCData_t** and shall not include the names **DataClass** or **DimensionalUnits**.
2. There are no required elements; all three lists may be empty.

This structure definition shows separate lists for global verses local data. The global data is essentially scalars, while the local data variables have size determined by the structure parameter **ListLength**. For **DataArray_t** entities with standardized data-name identifiers listed in Annex A, **DataType** is determined by convention. For user-defined variables, **DataType** is a user input.

Two important points need to be mentioned regarding this structure definition. First, this definition allows a given instance of **BCData_t** to have a mixture of global and local data. For example, if a user specifies Dirichlet data that has a uniform stagnation pressure but has a non-uniform velocity profile, this structure allows the user to describe the stagnation pressure by a scalar in the `DataGlobal` list and the velocity by an array in the `DataLocal` list. Second, the only distinction between the lists (aside from default names, which will be seldom used) is the parameters passed into the `DataArray_t` structure. Therefore, in actual implementation of this `BCData_t` structure it may not be possible to distinguish between members of the global and local lists without querying inside the `DataArray_t` substructures. Straightforward mapping onto the ADF database will not provide any distinctions between the members of the two lists. This hopefully will not cause any problems.

`DataClass` defines the default for the class of data contained in the boundary-condition data. If the boundary-condition data is dimensional, `DimensionalUnits` may be used to describe the system of dimensional units employed. If present, these two entities take precedence of all corresponding entities at higher levels of the hierarchy. These precedence rules are further discussed in Section 6.3.

9.6 Boundary Condition Type Structure Definition: `BCType_t`

`BCType_t` is an enumeration type that identifies the boundary-condition equations to be enforced at a given boundary location. `BCType_t` is a superset of two enumeration types, `BCTypeSimple_t` and `BCTypeCompound_t`.

```
BCTypeSimple_t := Enumeration(
    Null, BCGeneral, BCDirichlet, BCNeumann, BCExtrapolate, BCWallInviscid,
    BCWallViscousHeatFlux, BCWallViscousIsothermal, BCWallViscous, BCWall,
    BCInflowSubsonic, BCInflowSupersonic, BCOutflowSubsonic, BCOutflowSupersonic,
    BCTunnelInflow, BCTunnelOutflow, BCDegenerateLine, BCDegeneratePoint,
    BCSymmetryPlane, BCSymmetryPolar, BCAxisymmetricWedge, FamilySpecified,
    UserDefined ) ;


BCTypeCompound_t := Enumeration(
    Null, BCInflow, BCOutflow, BCFarfield, UserDefined ) ;
```

Any member of `BCTypeSimple_t` or `BCTypeCompound_t` is also a member of `BCType_t`. Simple boundary-condition types are described by `BCTypeSimple_t` and compound types by `BCTypeCompound_t`. Some members of `BCType_t` completely identify the equations to impose, while other give a general description of the class of boundary-condition equations to impose. The specific boundary-condition equations to enforce for each value of `BCType_t` are listed in Table 2 and Table 3.

The subdivision of `BCType_t` is based on function. For simple boundary conditions, the equations and data imposed are fixed; whereas, for compound boundary conditions different sets of equations are imposed depending on local flow conditions at the boundary. This distinction requires additional rules for dealing with simple and compound boundary-condition types. These rules are discussed in Section 9.7.

For the inflow/outflow boundary-condition descriptions, 3-D inviscid compressible flow is assumed; the 2-D equivalent should be obvious. These same boundary conditions are typically used for viscous cases also. This '3-D Euler' assumption will be noted wherever used.

In the following tables, Q is the solution vector, $\vec{q}$ is the velocity vector whose magnitude is q, the unit normal to the boundary is $\hat{n}$, and $\partial()/\partial n = \hat{n} \cdot \nabla$ is differentiation normal to the boundary.

Table 2: Simple Boundary-Condition Types

BCType_t or BCTypeSimple_t Identifier	Boundary-Condition Description
BCGeneral	Arbitrary conditions on Q or $\partial Q / \partial n$
BCDirichlet	Dirichlet condition on Q vector
BCNeumann	Neumann condition on $\partial Q / \partial n$
BCExtrapolate	Extrapolate Q from interior
BCWallInviscid	Inviscid (slip) wall • normal velocity specified (default: $\vec{q} \cdot \hat{n} = 0$)
BCWallViscousHeatFlux	Viscous no-slip wall with heat flux • velocity Dirichlet (default: $q = 0$) • temperature Neumann (default: adiabatic, $\partial T / \partial n = 0$)
BCWallViscousIsothermal	Viscous no-slip, isothermal wall • velocity Dirichlet (default: $q = 0$) • temperature Dirichlet
BCWallViscous	Viscous no-slip wall; special cases are BCWallViscousHeatFlux and BCWallViscousIsothermal • velocity Dirichlet (default: $q = 0$) • Dirichlet or Neumann on temperature
BCWall	General wall condition; special cases are BCWallInviscid, BCWallViscous, BCWallViscousHeatFlux and BCWallViscousIsothermal
BCInflowSubsonic	Inflow with subsonic normal velocity • specify 4; extrapolate 1 (3-D Euler)
BCInflowSupersonic	Inflow with supersonic normal velocity • specify 5; extrapolate 0 (3-D Euler) Same as BCDirichlet
BCOutflowSubsonic	Outflow with subsonic normal velocity • specify 1; extrapolate 4 (3-D Euler)
BCOutflowSupersonic	Outflow with supersonic normal velocity • specify 0; extrapolate 5 (3-D Euler) Same as BCExtrapolate

Continued on next page

Table 2: Simple Boundary-Condition Types (*Continued*)

BCType_t or BCTypeSimple_t Identifier	Boundary-Condition Description
BCTunnelInflow	Tunnel inlet (subsonic normal velocity) • specify cross-flow velocity, stagnation enthalpy, entropy • extrapolate 1 (3-D Euler)
BCTunnelOutflow	Tunnel exit (subsonic normal velocity) • specify static pressure • extrapolate 4 (3-D Euler)
BCDegenerateLine	Face degenerated to a line
BCDegeneratePoint	Face degenerated to a point
BCSymmetryPlane	Symmetry plane; face should be coplanar • density, pressure: $\partial()/\partial n = 0$ • tangential velocity: $\partial(\vec{q} \times \hat{n})/\partial n = 0$ • normal velocity: $\vec{q} \cdot \hat{n} = 0$
BCSymmetryPolar	Polar-coordinate singularity line; special case of BCDegenerateLine where degenerate face is a straight line and flowfield has polar symmetry; $\hat{s}$ is singularity line tangential unit vector • normal velocity: $\vec{q} \times \hat{s} = 0$ • all others: $\partial()/\partial n = 0$, n normal to $\hat{s}$
BCAxisymmetricWedge	Axisymmetric wedge; special case of BCDegenerateLine where degenerate face is a straight line
FamilySpecified	A boundary condition type is being specified for the family to which the current boundary belongs. A FamilyName_t specification must exist under BC_t, corresponding to a Family_t structure under CGNSBase_t. Under the Family_t structure there must be a FamilyBC_t structure specifying a valid BCType (other than FamilySpecified!). If any of these are absent, the boundary condition type is undefined.

Table 3: Compound Boundary-Condition Types

BCType_t or BCTypeCompound_t Identifier	Boundary-Condition Description
BCInflow	Inflow, arbitrary normal Mach; test on normal Mach, then perform one of: BCInflowSubsonic, BCInflowSupersonic
BCOutflow	Outflow, arbitrary normal Mach; test on normal Mach, then perform one of: BCOutflowSubsonic, BCOutflowSupersonic
BCFarfield	Farfield inflow/outflow, arbitrary normal Mach; test on normal velocity and normal Mach, then perform one of: BCInflowSubsonic, BCInflowSupersonic, BCOutflowSubsonic, BCOutflowSupersonic

9.7 Matching Boundary-Condition Data Sets

The BC_t structure allows for a arbitrary list of boundary-condition data sets, described by the BCDataSet_t structure. For simple boundary conditions, a single data set must be chosen from a list that may contain more than one element. Likewise, for a compound boundary condition, a limited number of data sets must be chosen and applied appropriately. The mechanism for proper choice of data sets is controlled by the BCType field of the BC_t structure, the BCTypeSimple field of the BCDataSet_t structure, and the boundary-condition type association table (Table 4). In the following discussion, we will use the '/' notation for fields or elements of a structure type.

BC_t is used for both simple and compound boundary conditions; hence, the field BC_t/BCType is of type BCType_t. Conversely, the substructure BCDataSet_t is intended to enforce a single set of boundary-condition equations independent of local flow conditions (i.e. it is appropriate only for simple boundary conditions). This is why the field BCDataSet_t/BCTypeSimple is of type BCTypeSimple_t and not BCType_t. The appropriate choice of data sets is determined by matching the field BC_t/BCType with the field BCDataSet_t/BCTypeSimple as specified in Table 4.

For simple boundary conditions, a single match from the list of BCDataSet_t instances is required. For all BCTypeSimple_t identifiers, except BCInflowSupersonic and BCOutflowSupersonic, an exact match is necessary. BCInflowSupersonic will match itself or BCDirichlet; BCOutflowSupersonic will match itself or BCExtrapolate.

For compound boundary conditions, the association table specifies which simple boundary-condition types are appropriate. Since compound boundary conditions enforce different boundary-condition equation sets depending on local flow conditions, several instances of BCDataSet_t will be matched for each BCTypeCompound_t identifier. The accompanying rule determines which of the matching data sets to apply at a given location on the BC patch.

This provides a general procedure applicable to both BCTypeSimple_t and BCTypeCompound_t situations. For a given BC_t/BCType use those instances of BCDataSet_t whose field BCDataSet_t/BCTypeSimple matches according to Table 4. Apply the matching data set or sets as prescribed by the appropriate usage rule.

Table 4: Associated Boundary-Condition Types and Usage Rules

BCType_t Identifier	Associated BCTypeSimple_t Identifiers and Usage Rules
BCInflow	BCInflowSupersonic BCInflowSubsonic *Usage Rule:* • if supersonic normal Mach, choose BCInflowSupersonic; • else, choose BCInflowSubsonic
BCOutflow	BCOutflowSupersonic BCOutflowSubsonic *Usage Rule:* • if supersonic normal Mach, choose BCOutflowSupersonic; • else, choose BCOutflowSubsonic
BCFarfield	BCInflowSupersonic BCInflowSubsonic BCOutflowSupersonic BCOutflowSubsonic *Usage Rule:* • if inflow and supersonic normal Mach, choose BCInflowSupersonic; • else if inflow, choose BCInflowSubsonic; • else if outflow and supersonic normal Mach, choose BCOutflowSupersonic; • else, choose BCOutflowSubsonic
BCInflowSupersonic	BCInflowSupersonic BCDirichlet *Usage Rule:* • choose either; BCInflowSupersonic takes precedence
BCOutflowSupersonic	BCOutflowSupersonic BCExtrapolate *Usage Rule:* • choose either; BCOutflowSupersonic takes precedence
All others	Self-matching

Although we present a strict division between the two categories of boundary-condition types, we realize that some overlap may exist. For example, some of the more general simple boundary-condition types, such as BCWall, may include a situation of inflow/outflow (say if the wall is porous). These complications require further guidelines on appropriate definition and use of boundary-condition types. The real distinctions between BCTypeSimple_t and BCTypeCompound_t are as follows:

- `BCTypeSimple_t` identifiers always match themselves; `BCTypeCompound_t` identifiers never match themselves.

- `BCTypeSimple_t` identifiers always produce a single match; `BCTypeCompound_t` will produce multiple matches.

- The usage rule for `BCTypeSimple_t` identifiers is always trivial—apply the single matching data set regardless of local flow conditions.

Therefore, any boundary condition that involves application of different data sets depending on local flow conditions should be classified `BCTypeCompound_t`. If a type that we have classified `BCTypeSimple_t` is used as a compound type (`BCWall` for a porous wall is an example), then it should somehow be reclassified. One option is to define a new `BCTypeCompound_t` identifier and provide associated `BCTypeSimple_t` types and a usage rule. Another option may be to allow some identifiers to be both `BCTypeSimple_t` and `BCTypeCompound_t` and let their appropriate use be based on context. This is still undetermined.

9.8 Boundary Condition Specification Data

For a given simple boundary condition (i.e. one that is not dependent on local flow conditions), the database provides a set of boundary-condition equations to be enforced through the structure definitions for `BCDataSet_t` and `BCData_t` (Section 9.4 and Section 9.5). Apart from the boundary-condition type, the precise equations to be enforced are described by boundary-condition solution data. These specified solution data are arranged by 'equation type':

Dirichlet: $Q = (Q)_{\text{specified}}$

Neumann: $\partial Q/\partial n = (\partial Q/\partial n)_{\text{specified}}$

The `DirichletData` and `NeumannData` entities of `BCData_t` list both the solution variables involved in the equations (through the data-name identifier conventions of Annex A) and the specified solution data.

Two issues need to be addressed for specifying Dirichlet or Neumann boundary-condition data. The first is whether the data is global or local:

Global BC data: Data applied globally to the BC patch; for example, specifying a uniform total pressure at an inflow boundary

Local BC data: Data applied locally at each vertex or cell face of the BC patch; an example of this is varying total pressure specified at each grid point at an inflow boundary

The second issue is describing the actual solution quantities that are to be specified. Both of these issues are addressed by use of the `DataArray_t` structure.

For some types of boundary conditions, many different combinations of solution quantities could be specified. For example, BCInflowSubsonic requires 4 solution quantities to be specified in 3-D, but what those 4 quantities are varies with applications (e.g. internal verses external flows) and codes. We propose the convention that the actual data being specified for any BCType is given by the list of DataArray_t entities included in DirichletData and NeumannData structures (actually by the identifier attached to each instance of DataArray_t). This frees us from having to define *many* versions of a given BCType (e.g. BCInflowSubsonic1, BCInflowSubsonic2, etc.), where each has a precisely defined set of Dirichlet data. We are left with the easier task of defining *how many* Dirichlet or Neumann quantities must be provided for each BCType.

An example of using DataArray_t-identifier conventions to describe BC specification data is the following: subsonic inflow with uniform stagnation pressure, mass flow and cross-flow angle specified; the Dirichlet data are stagnation pressure = 2.56, mass flow = 1.34, and cross-flow angle has a y-component of 0.043 and a z-component of 0.02 (ignore dimensional-units or normalization for the present). The specified solution variables and associated data are described as shown:

```
BCData_t<ListLength=?> DirichletData =
  {{
  DataArray_t<real, 1, 1> PressureStagnation = {{ Data(real, 1, 1) = 2.56  }} ;
  DataArray_t<real, 1, 1> MassFlow           = {{ Data(real, 1, 1) = 1.34  }} ;
  DataArray_t<real, 1, 1> VelocityAngleY     = {{ Data(real, 1, 1) = 0.043 }} ;
  DataArray_t<real, 1, 1> VelocityAngleZ     = {{ Data(real, 1, 1) = 0.02  }} ;
  }} ;
```

Basically, this states that DirichletData contains four instances of DataArray_t with identifiers or names PressureStagnation, MassFlow, VelocityAngleY and VelocityAngleZ. Each DataArray_t structure entity contains a single floating-point value; these are the Dirichlet data for the BC. Note that Data(real, 1, 1) means a single floating-point value.

The global verses local data issue can be easily handled by storing either a scalar, as shown above, for the global BC data case; or storing an array for the local BC data case. Storing an array of local BC data allows the capability for specifying non-constant solution profiles, such as 'analytic' boundary-layer profiles or profiles derived from experimental data. For the above example, if the stagnation pressure is instead specified at every vertex of the boundary-condition patch the following results:

```
BCData_t<ListLength=99> DirichletData =
  {{
  DataArray_t<real, 1, 99> PressureStagnation =
    {{ Data(real, 1, 99) = (PTOT(n), n=1,99) }} ;
  DataArray_t<real, 1, 1> MassFlow            = {{ Data(real, 1, 1) = 1.34  }} ;
  DataArray_t<real, 1, 1> VelocityAngleY      = {{ Data(real, 1, 1) = 0.043 }} ;
  DataArray_t<real, 1, 1> VelocityAngleZ      = {{ Data(real, 1, 1) = 0.02  }} ;
  }} ;
```

where, say, the boundary face is logically rectangular and contains 11×9 vertices and the stagnation pressure at the vertices is given by the array PTOT().

To facilitate implementation of boundary conditions into existing flow solvers, we adopt the convention that if no boundary-condition data is specified, then flow solvers are free to enforce any appropriate boundary-condition equations. This includes situations where entities of `BCDataSet_t`, `BCData_t` or `DataArray_t` are absent within the boundary-condition hierarchy. By convention, if no `BCDataSet` entities are present, then application codes are free to enforce appropriate BCs for the given value of `BCType`. Furthermore, if the entities `DirichletData` and `NeumannData` are not present in an instance of `BCDataSet_t`, or if insufficient data is present in `DirichletData` or `NeumannData` (e.g. if only one Dirichlet variable is present for a subsonic inflow condition), then application codes are free to fill out the boundary-condition data as appropriate for the `BCTypeSimple` identifier.

The various levels of BC implementation allowed are shown in Figure 5, from the lowest level in which the application codes interpret the `BCType`, to the fully SIDS-compliant BC implementation which completely defines the BC within the CGNS file.

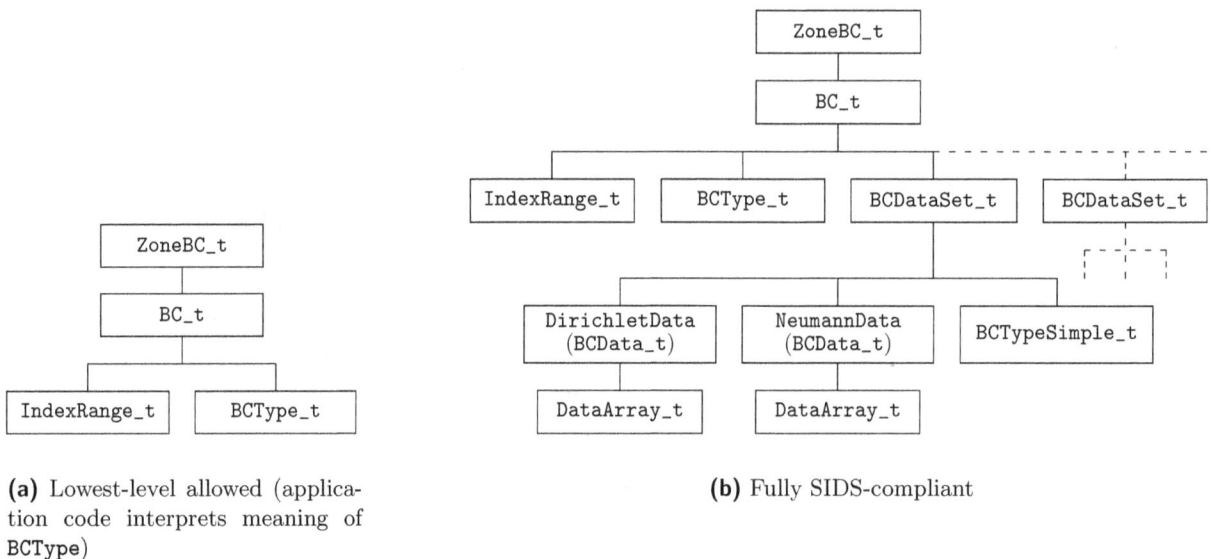

(a) Lowest-level allowed (application code interprets meaning of `BCType`)

(b) Fully SIDS-compliant

Figure 5: Boundary Condition Implementation Levels

An alternative approach to the present design could be to list all the solution variables and data (as `DataArray_t`-like structures) for the boundary condition, and contain descriptive tags in each one to indicate if they are Dirichlet or Neumann data. We have not taken this approach. We think grouping boundary-condition data by 'equation type' as we have done better allows for future extension to other types of boundary conditions (e.g. 2nd-order non-reflecting BC's that result in P.D.E.'s to be solved at the boundary).

9.9 Boundary Condition Examples

This section contains boundary-condition examples with increasing complexity. Included is the most simple `BC_t` entity and one of the most complex. The examples show situations of local and global boundary-condition data, simple and compound boundary-condition types, and multiple boundary-condition data sets that must be matched with the appropriate boundary-condition type.

Example 9-A: Symmetry Plane

Symmetry plane for a patch on the i-min face of a 3-D structured zone.

```
!  IndexDimension = 3
BC_t<3,3> BC1 =
  {{
  BCType_t BCType = BCSymmetryPlane ;

  IndexRange_t<3> PointRange =
    {{
    int[3] Begin = [1,1,1 ] ;
    int[3] End   = [1,9,17] ;
    }} ;
  }} ;
```

Since the boundary-condition equations to be enforced are completely defined by the boundary-condition type BCSymmetryPlane, no other information needs to be provided, except for the extent of the BC patch. The BC patch is specified by PointRange with a beginning index of (1,1,1) and an ending index of (1,9,17). By default, these refer to vertices.

Example 9-B: Viscous Solid Wall

A viscous solid wall for a 3-D structured zone, where a Dirichlet condition is enforced for temperature; the wall temperature for the entire wall is specified to be 273 K. The BC patch is on the j-min face and is bounded by the indices (1,1,1) and (33,1,9).

```
!  IndexDimension = 3
BC_t<3,3> BC2 =
  {{
  BCType_t BCType = BCWallViscousIsothermal ;

  IndexRange_t<3> PointRange =
    {{
    int[3] Begin = [1 ,1,1] ;
    int[3] End   = [33,1,9] ;
    }} ;

  !  ListLength = 33*9 = 297
  BCDataSet_t<297> BCDataSet1 =
    {{
    BCTypeSimple_t BCTypeSimple = BCWallViscousIsothermal ;

    !  Data array length = ListLength = 297
```

```
  BCData_t<297> DirichletData =
    {{
    DataArray_t<real, 1, 1> Temperature =
      {{
      Data(real, 1, 1) = 273. ;

      DataClass_t DataClass = Dimensional ;

      DimensionalUnits_t DimensionalUnits =
        {{
        MassUnits         = Null ;
        LengthUnits       = Null ;
        TimeUnits         = Null ;
        TemperatureUnits  = Kelvin ;
        AngleUnits        = Null ;
        }} ;
      }} ;
    }} ;
  }} ;
}} ;
```

This is an example of a simple boundary-condition type, BCWallViscousIsothermal. By default there is a zero Dirichlet condition on the velocity, and BCDataSet1 states there is a Dirichlet condition on temperature with a global value of 273 K. The data set contains a single BCData_t entity, called DirichletData, meaning a (possibly empty) collection of Dirichlet conditions should be enforced. Within DirichletData, there is a single DataArray_t entity; this narrows the specification to a single Dirichlet condition. This lone entity has the identifier Temperature, which by conventions defined in Annex A is the identifier for static temperature. The data contained in Temperature is a floating-point scalar with a value of 273. The qualifiers DataClass and DimensionalUnits specifies that the temperature is dimensional with units of Kelvin.

Since BCWallViscousIsothermal is a simple boundary-condition type, the appropriate data set contains a BCTypeSimple entity whose value is BCWallViscousIsothermal. For this example, only a single data set is provided, and this data set has the correct boundary-condition type. This is an example of a trivial data-set match.

Apart from velocity and temperature, additional 'numerical' boundary conditions are typically required by Navier-Stokes flow solvers, but none are given here; therefore, a code is free to implement other additional boundary conditions as desired.

Although the boundary-condition data is global, we include in this example structure parameters that are the lengths of potential local-data arrays. Comments are added to the example with the '!' notation to document the structure parameters. The BCDataSet_t structure function ListLength is evaluated based on PointRange. Since GridLocation is not specified in BC2, any local data is at vertices by default. The entity Temperature contains global data, so the value of ListLength is unused in DirichletData.

This example raises the question of whether unused structure parameters are required in structure entities. The answer is no. We included them here for completeness. The purpose of structure parameters is to mimic the need to define elements of a entity based on information contained elsewhere (at a higher level) in the CGNS database. When this need is not present in a given instance of a structure entity, the structure parameters are superfluous. Structure parameters that are superfluous or otherwise not needed in the following examples are denoted by '?'.

Example 9-C: Subsonic Inflow

Subsonic inflow for a 2-D structured zone: The BC patch is on the i-min face and includes $j \in [2, 7]$. As prescribed by the boundary-condition type, three quantities must be specified. Uniform entropy and stagnation enthalpy are specified with values of 0.94 and 2.85, respectively. A velocity profile is specified at face midpoints, given by the array **v_inflow(j)**. No dimensional or nondimensional information is provided.

```
!  IndexDimension = 2
BC_t<2,?> BC3 =
  {{
  BCType_t BCType = BCInflowSubsonic ;

  GridLocation_t GridLocation = FaceCenter ;

  IndexRange_t<2> PointRange =
    {{
    int[2] Begin = [1,2] ;
    int[2] End   = [1,6] ;
    }} ;

  !  ListLength = 5
  BCDataSet_t<5> BCDataSet1 =
    {{
    BCTypeSimple_t BCTypeSimple = BCInflowSubsonic ;

    !  Data array length = ListLength = 5
    BCData_t<5> DirichletData =
      {{
      DataArray_t<real, 1, 1> EntropyApprox =
        {{
        Data(real, 1, 1) = 0.94 ;
        }} ;

      DataArray_t<real, 1, 1> EnthalpyStagnation =
        {{
        Data(real, 1, 1) = 2.85 ;
        }} ;
```

```
      DataArray_t<real, 1, 5> VelocityY =
        {{
        Data(real, 1, 5) = (v_inflow(j), j=3,7) ;
        }} ;
      }} ;
    }} ;
  }} ;
```

This is another example of a simple boundary-condition type. The primary additional complexity included in this example is multiple Dirichlet conditions with one containing local data. DirichletData contains three DataArray_t entities named EntropyApprox, EnthalpyStagnation and VelocityY. This specifies three Dirichlet boundary conditions to be enforced, and the names identify the solution quantities to set. Since both EntropyApprox and EnthalpyStagnation have an array-length structure parameter of one, they identify global data, and the values are provided. VelocityY is an array of data values and contains the values in v_inflow(). The length of the array is given by ListLength, which represents the number of cell faces because BC3 contains the entity GridLocation whose value is FaceCenter. Note that the beginning and ending indices on the array v_inflow() are unimportant (they are user inputs); there just needs to be five values provided.

Example 9-D: Outflow

Outflow boundary condition with unspecified normal Mach number for an *i*-max face of a 3-D structured zone: for subsonic outflow, a uniform pressure is specified; for supersonic outflow, no boundary-condition equations are specified.

```
  !  IndexDimension = 3
  BC_t<3,3> BC4 =
    {{
    BCType_t BCType = BCOutflow ;

    IndexRange_t<3> PointRange = {{ }} ;

    BCDataSet_t<?> BCDataSetSubsonic =
      {{
      BCTypeSimple_t BCTypeSimple = BCOutflowSubsonic ;

      BCData_t<?> DirichletData =
        {{
        DataArray_t<real, 1, 1> Pressure = {{ }} ;
        }} ;
      }} ;

    BCDataSet_t<?> BCDataSetSupersonic =
      {{
      BCTypeSimple_t BCTypeSimple = BCOutflowSupersonic ;
```

```
    }} ;
  }} ;
```

This is an example of a complex boundary-condition type; the equation set to be enforced depends on the local flow conditions, namely the Mach number normal to the boundary. Two data sets are provided, `BCDataSetSubsonic` and `BCDataSetSupersonic`; recall the names are unimportant and are user defined. The first data set has a boundary-condition type of `BCOutflowSubsonic` and prescribes a global Dirichlet condition on static pressure. Any additional boundary conditions needed may be applied by a flow solver. The second data set has a boundary-condition type of `BCOutflowSupersonic` with no additional boundary-condition equation specification. Typically, all solution quantities are extrapolated from the interior for supersonic outflow. From the boundary-condition type association table (Table 4), `BCOutflow` requires two data sets with boundary-condition types `BCOutflowSubsonic` and `BCOutflowSupersonic`. The accompanying usage rule states that the data set for `BCOutflowSubsonic` should be used for a subsonic normal Mach number; otherwise, the data set for `BCOutflowSupersonic` should be enforced.

Any additional data sets with boundary-condition types other than `BCOutflowSubsonic` or `BCOutflowSupersonic` could be provided (the definition of `BC_t` allows an arbitrary list of `BCDataSet_t` entities); however, they should be ignored by any code processing the boundary-condition information. Another caveat is that providing two data sets with the same simple boundary-condition type would cause indeterminate results — which one is the correct data set to apply?

The actual global data value for static pressure is not provided; an abbreviated form of the `Pressure` entity is shown. This example also uses the '?' notation for unused data-array-length structure parameters.

Example 9-E: Farfield

Farfield boundary condition with arbitrary flow conditions for a j-max face of a 2-D structured zone: If subsonic inflow, specify entropy, vorticity and incoming acoustic characteristics; if supersonic inflow specify entire flow state; if subsonic outflow, specify incoming acoustic characteristic; and if supersonic outflow, extrapolate all flow quantities. None of the extrapolated quantities for the different boundary condition possibilities need be stated.

```
BC_t<2,2> BC5 =
  {{
  BCType_t BCType = BCFarfield ;

  IndexRange_t<2> PointRange = {{ }} ;

  int[2] InwardNormalIndex = [0,-1] ;

  BCDataSet_t<?> BCDataSetInflowSupersonic =
    {{
    BCTypeSimple_t BCTypeSimple = BCInflowSupersonic ;
    }} ;
```

```
BCDataSet_t<?> BCDataSetInflowSubsonic =
  {{
  BCTypeSimple_t BCTypeSimple = BCInflowSubsonic ;

  BCData<?> DirichletData =
    {{
    DataArray_t<real, 1, 1> CharacteristicEntropy     = {{ }} ;
    DataArray_t<real, 1, 1> CharacteristicVorticity1   = {{ }} ;
    DataArray_t<real, 1, 1> CharacteristicAcousticPlus = {{ }} ;
    }} ;
  }} ;

BCDataSet_t<?> BCDataSetOutflowSupersonic =
  {{
  BCTypeSimple_t BCTypeSimple = BCOutflowSupersonic ;
  }} ;

BCDataSet_t<?> BCDataSetOutflowSubsonic =
  {{
  BCTypeSimple_t BCTypeSimple = BCOutflowSubsonic ;

  BCData<?> DirichletData =
    {{
    DataArray_t<real, 1, 1> CharacteristicAcousticMinus = {{ }} ;
    }} ;
  }} ;
}} ;
```

The farfield boundary-condition type is the most complex of the compound boundary-condition types. BCFarfield requires four data sets; these data sets must contain the simple boundary-condition types BCInflowSupersonic, BCInflowSubsonic, BCOutflowSupersonic and BCOutflow-Subsonic. This example provides four appropriate data sets. The usage rule given for BCFarfield in Table 4 states which set of boundary-condition equations to be enforced based on the normal velocity and normal Mach number.

The data set for supersonic-inflow provides no information other than the boundary-condition type. A flow solver is free to apply any conditions that are appropriate; typically all solution quantities are set to freestream reference state values. The data set for subsonic-inflow states that three Dirichlet conditions should be enforced; the three data identifiers provided are among the list of conventions given in Annex A.5. The data set for supersonic-outflow only provides the boundary-condition type, and the data set for subsonic-outflow provides one Dirichlet condition on the incoming acoustic characteristic, CharacteristicAcousticMinus.

Also provided in the example is the inward-pointing computational-coordinate normal; the normal points in the $-j$ direction, meaning the BC patch is a j-max face. This information could also be obtained from the BC patch description given in IndexRange.

Note that this example shows only the overall layout of the boundary-condition entity. `IndexRange` and all `DataArray_t` entities are abbreviated, and all unused structure functions are not evaluated.

10   Governing Flow Equations

This section provides structure type definitions for describing the governing flow-equation set associated with the database. The description includes the general class of governing equations, the turbulent closure equations, the gas model, and the viscosity and thermal-conductivity models. Included with each equation description are associated constants. The structure definitions attempt to balance the opposing requirements for future growth and extensibility with initial ease of implementation. Included in the final section (Section 10.7) are examples of flow-equation sets.

The intended use of these structures initially is primarily for archival purposes and to provide additional documentation of the flow solution. If successful in this role, it is foreseeable that these flow-equation structures may eventually be also used as inputs for grid generators, flow solvers, and post-processors.

10.1   Flow Equation Set Structure Definition: `FlowEquationSet_t`

`FlowEquationSet_t` is a general description of the governing flow equations. It includes the dimensionality of the governing equations, and the collection of specific equation-set descriptions covered in subsequent sections. It can be a child node of either `CGNSBase_t` or `Zone_t` (or both).

```
FlowEquationSet_t< int CellDimension > :=
  {
  List( Descriptor_t Descriptor1 ... DescriptorN ) ;                    (o)

  int EquationDimension ;                                               (o)

  GoverningEquations_t<CellDimension> GoverningEquations ;              (o)

  GasModel_t GasModel ;                                                 (o)

  ViscosityModel_t ViscosityModel ;                                     (o)

  ThermalConductivityModel_t ThermalConductivityModel ;                 (o)

  TurbulenceClosure_t TurbulenceClosure ;                               (o)

  TurbulenceModel_t<CellDimension> TurbulenceModel ;                    (o)

  DataClass_t DataClass ;                                               (o)

  DimensionalUnits_t DimensionalUnits ;                                 (o)
  } ;
```

Notes

1. Default names for the `Descriptor_t` list are as shown; users may choose other legitimate names. Legitimate names must be unique within a given instance of `FlowEquationSet_t` and shall not include the names `EquationDimension`, `GoverningEquations`, `TurbulenceClosure`, `TurbulenceModel`, `GasModel`, `ViscosityModel`, `ThermalConductivityModel`, `DataClass` or `DimensionalUnits`.
2. There are no required elements for `FlowEquationSet_t`.

`FlowEquationSet_t` requires a single structure parameter, `CellDimension`, to identify the dimensionality of index arrays for structured grids. This parameter is passed onto several substructures.

`EquationDimension` is the dimensionality of the governing equations; it is the number of spatial variables describing the flow. `GoverningEquations` describes the general class of flow equations. `GasModel` describes the equation of state, and `ViscosityModel` and `ThermalConductivityModel` describe the auxiliary relations for molecular viscosity and the thermal conductivity coefficient. `TurbulenceClosure` and `TurbulenceModel` describe the turbulent closure for the Reynolds-averaged Navier-Stokes equations.

`DataClass` defines the default for the class of data contained in the flow-equation set. For any data that is dimensional, `DimensionalUnits` may be used to describe the system of dimensional units employed. If present, these two entities take precedence of all corresponding entities at higher levels of the hierarchy. These precedence rules are further discussed in Section 6.3.

10.2 Governing Equations Structure Definition: `GoverningEquations_t`

`GoverningEquations_t` describes the class of governing flow equations associated with the solution.

```
GoverningEquationsType_t := Enumeration(
  Null,
  FullPotential,
  Euler,
  NSLaminar,
  NSTurbulent,
  NSLaminarIncompressible,
  NSTurbulentIncompressible,
  UserDefined ) ;

GoverningEquations_t< int CellDimension > :=
  {
  List( Descriptor_t Descriptor1 ... DescriptorN ) ;              (o)

  GoverningEquationsType_t GoverningEquationsType ;               (r)

  int[CellDimension*(CellDimension + 1)/2] DiffusionModel ;       (o)
  } ;
```

Notes

1. Default names for the `Descriptor_t` list are as shown; users may choose other legitimate names. Legitimate names must be unique within a given instance of `GoverningEquations_t` and shall not include the name `DiffusionModel`.
2. `GoverningEquationsType` is the only required element.
3. The length of the `DiffusionModel` array is as follows: in 1-D it is `int[1]`; in 2-D it is `int[3]`; and in 3-D it is `int[6]`. For unstructured zones, `DiffusionModel` is not supported, and should not be used.

`GoverningEquations_t` requires a single structure parameter, `CellDimension`. It is used to define the length of the array `DiffusionModel`.

`DiffusionModel` describes the viscous diffusion terms modeled in the flow equations, and is applicable only to the Navier-Stokes equations with structured grids. Typically, thin-layer approximations include only the diffusion terms in one or two computational-coordinate directions. `Diffusion-Model` encodes the coordinate directions that include second-derivative and cross-derivative diffusion terms. The first `CellDimension` elements are second-derivative terms and the remainder elements are cross-derivative terms. Allowed values for individual elements in the array `DiffusionModel` are 0 and 1; a value of 1 indicates the diffusion term is modeled, and 0 indicates that they are not modeled. In 3-D, the encoding of `DiffusionModel` is as follows:

Element	Modeled Terms
$n = 1$	Diffusion terms in i ($\partial^2/\partial\xi^2$)
$n = 2$	Diffusion terms in j ($\partial^2/\partial\eta^2$)
$n = 3$	Diffusion terms in k ($\partial^2/\partial\zeta^2$)
$n = 4$	Cross-diffusion terms in i-j ($\partial^2/\partial\xi\partial\eta$ and $\partial^2/\partial\eta\partial\xi$)
$n = 5$	Cross-diffusion terms in j-k ($\partial^2/\partial\eta\partial\zeta$ and $\partial^2/\partial\zeta\partial\eta$)
$n = 6$	Cross-diffusion terms in k-i ($\partial^2/\partial\zeta\partial\xi$ and $\partial^2/\partial\xi\partial\zeta$)

where derivatives in the i, j and k computational-coordinates are ξ, η and ζ, respectively. The full Navier-Stokes equations in 3-D are indicated by `DiffusionModel = [1,1,1,1,1,1]`, and the thin-layer equations including only diffusion in the j-direction are `[0,1,0,0,0,0]`.

10.3 Thermodynamic Gas Model Structure Definition: `GasModel_t`

`GasModel_t` describes the equation of state model used in the governing equations to relate pressure, temperature and density.

```
GasModelType_t := Enumeration(
  Null,
  Ideal,
  VanderWaals,
  UserDefined ) ;
```

```
GasModel_t :=
  {
  List( Descriptor_t Descriptor1 ... DescriptorN ) ;                     (o)

  GasModelType_t GasModelType ;                                          (r)

  List( DataArray_t<DataType, 1, 1> DataArray1 ... DataArrayN ) ;        (o)

  DataClass_t DataClass ;                                                (o)

  DimensionalUnits_t DimensionalUnits ;                                  (o)
  } ;
```

Notes

1. Default names for the `Descriptor_t` and `DataArray_t` lists are as shown; users may choose other legitimate names. Legitimate names must be unique within a given instance of `GasModel_t` and shall not include the names `DataClass` or `DimensionalUnits`.
2. `GasModelType` is the only required element.

For a perfect gas (`GasModelType = Ideal`), the pressure, temperature and density are related by,

$$p = \rho R T,$$

where R is the ideal gas constant. Related quantities are the specific heat at constant pressure (c_p), specific heat at constant volume (c_v) and specific heat ratio ($\gamma = c_p/c_v$). The gas constant and specific heats are related by $R = c_p - c_v$. Data-name identifiers associated with the perfect gas law are listed in Table 5.

Table 5: Data-Name Identifiers for Perfect Gas

Data-Name Identifier	Description	Units
IdealGasConstant	Ideal gas constant (R)	$\mathbf{L}^2/(\mathbf{T}^2\Theta)$
SpecificHeatRatio	Ratio of specific heats ($\gamma = c_p/c_v$)	-
SpecificHeatVolume	Specific heat at constant volume (c_v)	$\mathbf{L}^2/(\mathbf{T}^2\Theta)$
SpecificHeatPressure	Specific heat at constant pressure (c_p)	$\mathbf{L}^2/(\mathbf{T}^2\Theta)$

If it is desired to specify any of these identifiers in a CGNS database, they should be defined as `DataArrays` under `GasModel_t`.

The dimensional units are defined as follows: $\mathbf{M}$ is mass, $\mathbf{L}$ is length, $\mathbf{T}$ is time and Θ is temperature. These are further described in Annex A.

`DataClass` defines the default for the class of data contained in the thermodynamic gas model. For any data that is dimensional, `DimensionalUnits` may be used to describe the system of dimensional units employed. If present, these two entities take precedence of all corresponding entities at higher levels of the hierarchy. These precedence rules are further discussed in Section 6.3.

10.4 Molecular Viscosity Model Structure Definition: `ViscosityModel_t`

`ViscosityModel_t` describes the model for relating molecular viscosity (μ) to temperature.

```
ViscosityModelType_t := Enumeration(
  Null,
  Constant,
  PowerLaw,
  SutherlandLaw,
  UserDefined ) ;

ViscosityModel_t :=
  {
  List( Descriptor_t Descriptor1 ... DescriptorN ) ;                    (o)

  ViscosityModelType_t ViscosityModelType ;                            (r)

  List( DataArray_t<DataType, 1, 1> DataArray1 ... DataArrayN ) ;      (o)

  DataClass_t DataClass ;                                              (o)

  DimensionalUnits_t DimensionalUnits ;                                (o)
  } ;
```

Notes

1. Default names for the `Descriptor_t` and `DataArray_t` lists are as shown; users may choose other legitimate names. Legitimate names must be unique within a given instance of `ViscosityModel_t` and shall not include the names `DataClass` or `DimensionalUnits`.
2. `ViscosityModelType` is the only required element.

The molecular viscosity models are as follows: `Constant` states that molecular viscosity is constant throughout the field and is equal to some reference value ($\mu = \mu_{\text{ref}}$); `PowerLaw` states that molecular viscosity follows a power-law relation,

$$\mu = \mu_{\text{ref}} \left(\frac{T}{T_{\text{ref}}} \right)^n$$

and `SutherlandLaw` is Sutherland's Law for molecular viscosity,

$$\mu = \mu_{\text{ref}} \left(\frac{T}{T_{\text{ref}}} \right)^{3/2} \frac{T_{\text{ref}} + T_s}{T + T_s},$$

where T_s is the Sutherland's Law constant, and μ_{ref} and T_{ref} are the reference viscosity and temperature, respectively. For air[2], the power-law exponent is $n = 0.666$, Sutherland's law constant (T_s) is 110.6 K, the reference temperature (T_{ref}) is 273.15 K, and the reference viscosity (μ_{ref})

Table 6: Data-Name Identifiers for Molecular Viscosity Models

ViscosityModelType	Data-Name Identifer	Description	Units
PowerLaw	PowerLawExponent	Power-law exponent (n)	-
SutherlandLaw	SutherlandLawConstant	Sutherland's Law constant (T_s)	Θ
All	TemperatureReference	Reference temperature (T_{ref})	Θ
All	ViscosityMolecularReference	Reference viscosity (μ_{ref})	$\mathbf{M/(LT)}$

is 1.716×10^{-5} kg/(m-s). The data-name identifiers for molecular viscosity models are defined in Table 6.

If it is desired to specify any of these identifiers in a CGNS database, they should be defined as DataArrays under ViscosityModel_t.

DataClass defines the default for the class of data contained in the molecular viscosity model. For any data that is dimensional, DimensionalUnits may be used to describe the system of dimensional units employed. If present, these two entities take precedence of all corresponding entities at higher levels of the hierarchy. These precedence rules are further discussed in Section 6.3.

10.5 Thermal Conductivity Model Structure Definition: ThermalConductivityModel_t

ThermalConductivityModel_t describes the model for relating the thermal-conductivity coefficient (k) to temperature.

```
ThermalConductivityModelType_t := Enumeration(
  Null,
  ConstantPrandtl,
  PowerLaw,
  SutherlandLaw,
  UserDefined ) ;

ThermalConductivityModel_t :=
  {
  List( Descriptor_t Descriptor1 ... DescriptorN ) ;          (o)

  ThermalConductivityModelType_t ThermalConductivityModelType ;  (r)

  List( DataArray_t<DataType, 1, 1> DataArray1 ... DataArrayN ) ;  (o)

  DataClass_t DataClass ;                                      (o)
```

[2]White, F. M., *Viscous Fluid Flow*, McGraw-Hill, 1974, p. 28-29

```
    DimensionalUnits_t DimensionalUnits ;                                    (o)
    } ;
```

Notes

1. Default names for the `Descriptor_t` and `DataArray_t` lists are as shown; users may choose other legitimate names. Legitimate names must be unique within a given instance of `Thermal-ConductivityModel_t` and shall not include the names `DataClass` or `DimensionalUnits`.
2. `ThermalConductivityModelType` is the only required element.

The thermal-conductivity models parallel the molecular viscosity models. `ConstantPrandtl` states that the Prandtl number ($Pr = \mu c_p / k$) is constant and equal to some reference value. `PowerLaw` relates k to temperature via a power-law relation,

$$k = k_{\text{ref}} \left(\frac{T}{T_{\text{ref}}} \right)^n .$$

`SutherlandLaw` states the Sutherland's Law for thermal conductivity,

$$k = k_{\text{ref}} \left(\frac{T}{T_{\text{ref}}} \right)^{3/2} \frac{T_{\text{ref}} + T_s}{T + T_s},$$

where k_{ref} is the reference thermal conductivity, T_{ref} is the reference temperature, and T_s is the Sutherland's law constant. For air[3], the Prandtl number is $Pr = 0.72$, the power-law exponent is $n = 0.81$, Sutherland's law constant (T_s) is 194.4 K, the reference temperature (T_{ref}) is 273.15 K, and the reference thermal conductivity (k_{ref}) is 2.414×10^{-2} kg-m/(s³-K). Data-name identifiers for thermal conductivity models are listed in Table 7.

Table 7: Data-Name Identifiers for Thermal Conductivity Models

ThermalConduc-tivityModelType	Data-Name Identifer	Description	Units
ConstantPrandtl	Prandtl	Prandtl number (Pr)	-
PowerLaw	PowerLawExponent	Power-law exponent (n)	-
SutherlandLaw	SutherlandLawConstant	Sutherland's Law constant (T_s)	Θ
All	TemperatureReference	Reference temperature (T_{ref})	Θ
All	ThermalConductivityReference	Reference thermal conductivity (k_{ref})	$\mathbf{ML/(T^3\Theta)}$

If it is desired to specify any of these identifiers in a CGNS database, they should be defined as `DataArrays` under `ThermalConductivityModel_t`.

---

[3]White, F. M., *Viscous Fluid Flow*, McGraw-Hill, 1974, p. 32-33

`DataClass` defines the default for the class of data contained in the thermal conductivity model. For any data that is dimensional, `DimensionalUnits` may be used to describe the system of dimensional units employed. If present, these two entities take precedence of all corresponding entities at higher levels of the hierarchy. These precedence rules are further discussed in Section 6.3.

10.6 Turbulence Structure Definitions

This section presents structure definitions for describing the form of closure used in the Reynolds-averaged (or Favre-averaged) Navier-Stokes equations for determining the Reynolds stress terms. Here 'turbulence closure' refers to eddy viscosity or other approximations for the Reynolds stress terms, and 'turbulence model' refers to the actual algebraic or turbulence-transport equation models used. To an extent these are independent choices (e.g. using either an eddy viscosity closure or an algebraic Reynolds-stress closure with a two-equation model).

10.6.1 Turbulence Closure Structure Definition: `TurbulenceClosure_t`

`TurbulenceClosure_t` describes the turbulence closure for the Reynolds stress terms of the Navier-Stokes equations.

```
TurbulenceClosureType_t := Enumeration(
  Null,
  EddyViscosity,
  ReynoldsStress,
  ReynoldsStressAlgebraic,
  UserDefined ) ;

TurbulenceClosure_t :=
  {
  List( Descriptor_t Descriptor1 ... DescriptorN ) ;              (o)

  TurbulenceClosureType_t TurbulenceClosureType ;                 (r)

  List( DataArray_t<DataType, 1, 1> DataArray1 ... DataArrayN ) ; (o)

  DataClass_t DataClass ;                                         (o)

  DimensionalUnits_t DimensionalUnits ;                           (o)
  } ;
```

Notes

1. Default names for the `Descriptor_t` and `DataArray_t` lists are as shown; users may choose other legitimate names. Legitimate names must be unique within a given instance of `TurbulenceClosure_t` and shall not include the names `DataClass` or `DimensionalUnits`.
2. `TurbulenceClosureType` is the only required element.

The different types of turbulent closure are as follows: `EddyViscosity` is the Boussinesq eddy-viscosity closure, where the Reynolds stresses are approximated as the product of an eddy viscosity (ν_t) and the mean strain tensor. Using indicial notation, the relation is,

$$-\overline{u_i' u_j'} = \nu_t \left(\frac{\partial u_i}{\partial x_j} + \frac{\partial u_j}{\partial x_i} \right),$$

where $-\overline{u_i' u_j'}$ are the Reynolds stresses; the notation is further discussed in Annex A.2. `ReynoldsStress` is no approximation of the Reynolds stresses. `ReynoldsStressAlgebraic` is an algebraic approximation for the Reynolds stresses based on some intermediate transport quantities.

Associated with the turbulent closure is a list of constants, where each constant is described by a separate `DataArray_t` entity. Constants associated with the eddy-viscosity closure are listed in Table 8.

Table 8: Data-Name Identifiers for Turbulence Closure

Data-Name Identifier	Description	Units
PrandtlTurbulent	Turbulent Prandtl number ($\rho\nu_t c_p / k_t$)	-

If it is desired to specify any of these identifiers in a CGNS database, they should be defined as `DataArrays` under `TurbulenceClosure_t`.

`DataClass` defines the default for the class of data contained in the turbulence closure. For any data that is dimensional, `DimensionalUnits` may be used to describe the system of dimensional units employed. If present, these two entities take precedence of all corresponding entities at higher levels of the hierarchy. These precedence rules are further discussed in Section 6.3.

10.6.2   Turbulence Model Structure Definition: `TurbulenceModel_t`

`TurbulenceModel_t` describes the equation set used to model the turbulence quantities.

```
  TurbulenceModelType_t := Enumeration(
    Null,
    Algebraic_BaldwinLomax,
    Algebraic_CebeciSmith,
    HalfEquation_JohnsonKing,
    OneEquation_BaldwinBarth,
    OneEquation_SpalartAllmaras,
    TwoEquation_JonesLaunder,
    TwoEquation_MenterSST,
    TwoEquation_Wilcox,
    UserDefined ) ;

  TurbulenceModel_t< int CellDimension > :=
```

```
{
   List( Descriptor_t Descriptor1 ... DescriptorN ) ;                    (o)

   TurbulenceModelType_t TurbulenceModelType ;                           (r)

   List( DataArray_t<DataType, 1, 1> DataArray1 ... DataArrayN ) ;       (o)

   int[CellDimension*(CellDimension + 1)/2] DiffusionModel ;             (o)

   DataClass_t DataClass ;                                               (o)

   DimensionalUnits_t DimensionalUnits ;                                 (o)
} ;
```

Notes

1. Default names for the `Descriptor_t` and `DataArray_t` lists are as shown; users may choose other legitimate names. Legitimate names must be unique within a given instance of `TurbulenceModel_t` and shall not include the names `DiffusionModel`, `DataClass`, or `DimensionalUnits`.
2. TurbulenceModelType is the only required element.
3. The length of the `DiffusionModel` array is as follows: in 1-D it is `int[1]`; in 2-D it is `int[3]`; and in 3-D it is `int[6]`. For unstructured zones, `DiffusionModel` is not supported, and should not be used.

`TurbulenceModel_t` requires a single structure parameter, `CellDimension`. It is used to define the length of the array `DiffusionModel`. `DiffusionModel` describes the viscous diffusion terms included in the turbulent transport model equations; the encoding of `DiffusionModel` is described in Section 10.2.

Associated with each choice of turbulence model may be a list of constants, where each constant is described by a separate `DataArray_t` entity. If used, the Data-Name Identifier of each constant should include the turbulence model name, as well as the constant name (e.g., `TurbulentSACb1`, `TurbulentSSTCmu`, `TurbulentKESigmak`, etc.). However, no attempt is made here to formalize the names for all possible turbulence models.

`DataClass` defines the default for the class of data contained in the turbulence model equation set. For any data that is dimensional, `DimensionalUnits` may be used to describe the system of dimensional units employed. If present, these two entities take precedence of all corresponding entities at higher levels of the hierarchy. These precedence rules are further discussed in Section 6.3.

Example 10-A: Spalart-Allmaras Turbulence Model

Description for the eddy-viscosity closure and Spalart-Allmaras turbulence model, including associated constants.

```
TurbulenceClosure_t TurbulenceClosure =
```

```
{{
TurbulenceClosureType_t TurbulenceClosureType = EddyViscosity ;

DataArray_t<real, 1, 1> PrandtlTurbulent = {{ 0.90 }} ;
}} ;

TurbulenceModel_t TurbulenceModel =
  {{
  TurbulenceModelType_t TurbulenceModelType = OneEquation_SpalartAllmaras ;

  DataArray_t<real, 1, 1> TurbulentSACb1   = {{ 0.1355 }} ;
  DataArray_t<real, 1, 1> TurbulentSACb2   = {{ 0.622 }} ;
  DataArray_t<real, 1, 1> TurbulentSASigma = {{ 2/3 }} ;
  DataArray_t<real, 1, 1> TurbulentSAKappa = {{ 0.41 }} ;
  DataArray_t<real, 1, 1> TurbulentSACw1   = {{ 3.2391 }} ;
  DataArray_t<real, 1, 1> TurbulentSACw2   = {{ 0.3 }} ;
  DataArray_t<real, 1, 1> TurbulentSACw3   = {{ 2 }} ;
  DataArray_t<real, 1, 1> TurbulentSACv1   = {{ 7.1 }} ;
  DataArray_t<real, 1, 1> TurbulentSACt1   = {{ 1 }} ;
  DataArray_t<real, 1, 1> TurbulentSACt2   = {{ 2 }} ;
  DataArray_t<real, 1, 1> TurbulentSACt3   = {{ 1.2 }} ;
  DataArray_t<real, 1, 1> TurbulentSACt4   = {{ 0.5 }} ;
  }} ;
```

Note that each **DataArray_t** entity is abbreviated.

10.7   Flow Equation Examples

This section presents two examples of flow-equation sets. The first is an inviscid case and the second is a turbulent case with a one-equation turbulence model.

Example 10-B: 3-D Compressible Euler

3-D compressible Euler with a perfect gas assumption for a monatomic gas:

```
FlowEquationSet_t<3> EulerEquations =
  {{
  int EquationDimension = 3 ;

  GoverningEquations_t<3> GoverningEquations =
    {{
    GoverningEquationsType_t GoverningEquationsType = Euler ;
    }} ;

  GasModel_t GasModel =
    {{
```

```
      GasModelType_t GasModelType = Ideal ;

      DataArray_t<real, 1, 1> SpecificHeatRatio =
        {{
        Data(real, 1, 1) = 1.667 ;

        DataClass_t DataClass = NondimensionalParameter ;
        }} ;
      }} ;
    }} ;
```

Example 10-C: 3-D Compressible Navier-Stokes

3-D compressible Navier-Stokes for a structured grid, with the S-A turbulence model, a perfect gas assumption, Sutherland's law for the molecular viscosity, a constant Prandtl-number assumption, and inclusion of the full Navier-Stokes diffusion terms; all models assume air:

```
  FlowEquationSet_t<3> NSEquations =
    {{
    int EquationDimension = 3 ;

    GoverningEquations_t<3> GoverningEquations =
      {{
      GoverningEquationsType_t GoverningEquationsType = NSTurbulent ;

      int[6] DiffusionModel = [1,1,1,1,1,1] ;
      }} ;

    GasModel_t GasModel =
      {{
      GasModelType_t GasModelType = Ideal ;

      DataArray_t<real, 1, 1> SpecificHeatRatio = {{ 1.4 }} ;
      }} ;

    ViscosityModel_t ViscosityModel =
      {{
      ViscosityModelType_t ViscosityModelType = SutherlandLaw ;

      DataArray_t<real, 1, 1> SutherlandLawConstant =
        {{
        Data(real, 1, 1) = 110.6 }} ;

        DataClass_t DataClass = Dimensional ;
        DimensionalUnits_t DimensionalUnits = {{ TemperatureUnits = Kelvin }} ;
```

```
    }} ;
  }} ;

  ThermalConductivityModel_t ThermalConductivityModel =
    {{
    ThermalConductivityModelType_t ThermalConductivityModelType =
      ConstantPrandtl ;

    DataArray_t<real, 1, 1> Prandtl = {{ 0.72 }} ;
    }} ;

  TurbulenceClosure_t<3> TurbulenceClosure =
    {{
    TurbulenceClosureType_t TurbulenceClosureType = EddyViscosity ;

    DataArray<real, 1, 1> PrandtlTurbulent = {{ 0.90 }} ;
    }} ;

  TurbulenceModel_t<3> TurbulenceModel =
    {{
    TurbulenceModelType_t TurbulenceModelType = OneEquation_SpalartAllmaras ;

    int[6] DiffusionModel = [1,1,1,1,1,1] ;
    }} ;
  }} ;
```

Note that all **DataArray_t** entities are abbreviated except **SutherlandLawConstant**.

11 Time-Dependent Flow

This section describes structure types intended primarily for time-dependent flows. Data structures are presented for storing time-dependent or iterative data, and for recording rigid and arbitary grid motion. The section concludes with several examples.

11.1 Iterative Data Structure Definitions

In order to keep a record of time dependent or iterative data, the data structures BaseIterative-Data_t and ZoneIterativeData_t are used.

11.1.1 Base Iterative Data Structure Definition: BaseIterativeData_t

The BaseIterativeData_t data structure is located directly under the CGNSBase_t node. It contains information about the number of time steps or iterations being recorded, and the time and/or iteration values at each step. In addition, it may include the list of zones and families for each step of the simulation, if these vary throughout the simulation.

The BaseIterativeData_t data structure is defined as follows:

```
BaseIterativeData_t :=
  {
  int NumberOfSteps                                              (r)

  DataArray_t<real, 1, NumberOfSteps> TimeValues ;              (o/r)
  DataArray_t<int,  1, NumberOfSteps> IterationValues ;         (r/o)

  DataArray_t<int,  1, NumberOfSteps> NumberOfZones ;           (o)
  DataArray_t<int,  1, NumberOfSteps> NumberOfFamilies ;        (o)
  DataArray_t<char, 3, [32, MaxNumberOfZones, NumberOfSteps]>
     ZonePointers ;                                             (o)
  DataArray_t<char, 3, [32, MaxNumberOfFamilies, NumberOfSteps]>
     FamilyPointers ;                                           (o)

  List( DataArray_t<> DataArray1 ... DataArrayN ) ;             (o)

  List( Descriptor_t Descriptor1 ... DescriptorN ) ;            (o)

  DataClass_t DataClass ;                                       (o)

  DimensionalUnits_t DimensionalUnits ;                         (o)
  }
```

Notes

1. NumberOfSteps is a required element of the BaseIterativeData_t data structure. It holds either the number of time steps or the number of iterations being recorded.
2. Either TimeValues or IterationValues must be defined. If both are used, there must be a one-to-one correspondence between them.

TimeValues and IterationValues are data-name identifiers corresponding to the time and iteration values stored in the file. When IterationValues are used, the iterative data stored in the database correspond to values at the end of the associated iteration.

The data-name identifiers NumberOfZones and ZonePointers are only used if different zone data structures apply to different steps of the simulation. (See Example 11-C.)

Similarly, if the geometry varies with time or iteration, then the data-name identifiers NumberOf-Families and FamilyPointers are used to record which Family_t data structure(s) correspond(s) to which step.

The DataArray_t nodes for ZonePointers and FamilyPointers are defined as three-dimensional arrays. For each recorded step, the names of all zones and families being used for the step may be recorded. Note that the names are limited to 32 characters, as usual in the SIDS. The variables MaxNumberOfZones and MaxNumberOfFamilies represent the maximum number of zones and families that apply to one step. So if NumberOfSteps = 5 and NumberOfZones = {2,2,3,4,3}, then MaxNumberOfZones equals 4.

When NumberOfZones and NumberOfFamilies vary for different stored time steps, the name Null is used in ZonePointers and FamilyPointers as appropriate for steps in which the NumberOfZones or NumberOfFamilies is less than the MaxNumberOfZones or MaxNumberOfFamilies.

Any number of extra DataArray_t nodes are allowed. These should be used to record data not covered by this specification.

11.1.2 Zone Iterative Data Structure Definition: ZoneIterativeData_t

The ZoneIterativeData_t data structure is located under the Zone_t node. It may be used to record pointers to zonal data for each recorded step of the simulation, and is defined as follows:

```
ZoneIterativeData_t< int NumberOfSteps > :=
  {
  DataArray_t<char, 2, [32, NumberOfSteps]> RigidGridMotionPointers ;      (o)
  DataArray_t<char, 2, [32, NumberOfSteps]> ArbitraryGridMotionPointers ; (o)
  DataArray_t<char, 2, [32, NumberOfSteps]> GridCoordinatesPointers ;      (o)
  DataArray_t<char, 2, [32, NumberOfSteps]> FlowSolutionsPointers ;        (o)

  List( DataArray_t<> DataArray1 ... DataArrayN ) ;                        (o)

  List( Descriptor_t Descriptor1 ... DescriptorN ) ;                      (o)
```

```
DataClass_t DataClass ;                                              (o)

DimensionalUnits_t DimensionalUnits ;                                (o)
}
```

The data arrays with data-name identifiers xxxPointers contain lists of associated data structures for each recorded time value or iteration. These data structures contain data at the associated time value, or at the end of the associated iteration. There is an implied one-to-one correspondence between each pointer (from 1, 2, ..., NumberOfSteps) and the associated TimeValues and/or IterationValues under BaseIterativeData_t. They refer by name to data structures within the current zone. The name Null is used when a particular time or iteration does not have a corresponding data structure to point to.

Any number of extra DataArray_t nodes are allowed. These should be used to record data not covered by this specification.

The ZoneIterativeData_t data structure may not exist without the BaseIterativeData_t under the CGNSBase_t node. However BaseIterativeData_t may exist without ZoneIterativeData_t.

11.2   Rigid Grid Motion Structure Definition: RigidGridMotion_t

Adding rigid grid motion information to the CGNS file enables an application code to determine the mesh location without the need to alter the original mesh definition recorded under GridCoordinates_t. A data structure named RigidGridMotion_t is used to record the necessary data defining a rigid translation and/or rotation of the grid coordinates.

The rigid grid motion is recorded independently for each zone of the CGNS base. Therefore the RigidGridMotion_t data structure is located under the zone data structure (Zone_t). There may be zero to several RigidGridMotion_t nodes under a Zone_t node. The multiple rigid grid motion definitions may be associated with different iterations or time steps in the computation. This association is recorded under the ZoneIterativeData_t data structure, described in Section 11.1.2.

```
RigidGridMotion_t :=
  {
  List( Descriptor_t Descriptor1 ... DescriptorN ) ;                 (o)

  RigidGridMotionType_t RigidGridMotionType ;                        (r)

  DataArray_t<real, 2, [PhysicalDimension, 2]> OriginLocation ;      (r)
  DataArray_t<real, 1,  PhysicalDimension>    RigidRotationAngle ;   (o/d)
  DataArray_t<real, 1,  PhysicalDimension>    RigidVelocity ;        (o)
  DataArray_t<real, 1,  PhysicalDimension>    RigidRotationRate ;    (o)

  List( DataArray_t DataArray1 ... DataArrayN ) ;                    (o)

  DataClass_t DataClass ;                                            (o)
```

```
    DimensionalUnits_t DimensionalUnits ;                              (o)
    } ;
```

Notes

1. `RigidGridMotionType` and `OriginLocation` are the only required elements under `Rigid-GridMotion_t`. All other elements are optional.

`RigidGridMotionType_t` is an enumeration type that describes the type of rigid grid motion.

```
  RigidGridMotionType_t := Enumeration(
    Null,
    ConstantRate,
    VariableRate,
    UserDefined ) ;
```

The characteristics of the grid motion are defined by the data-name identifiers in Table 9.

Table 9: Data-Name Identifiers for Rigid Grid Motion

Data-Name Identifier	Description	Units
OriginLocation	Physical coordinates of the origin before and after the rigid grid motion	**L**
RigidRotationAngle	Rotation angles about each axis of the translated coordinate system. If not specified, `RigidRotationAngle` is set to zero.	α
RigidVelocity	Grid velocity vector of the origin translation	**L/T**
RigidRotationRate	Rotation rate vector about the axis of the translated coordinate system	$\alpha/\mathbf{T}$

Any number of additional `DataArray_t` nodes are allowed. These may be used to record data not covered by this specification.

"Rigid grid motion" implies relative motion of grid zones. However, no attempt is made in the `RigidGridMotion_t` data structure to require that the `ZoneGridConnectivity_t` information be updated to be consistent with the new grid locations. Whether the `ZoneGridConnectivity_t` information refers to the original connectivity (of `GridCoordinates`) or the latest connectivity (of the moved or deformed grid) is currently left up to the user.

11.3 Arbitrary Grid Motion Structure Definition: `ArbitraryGridMotion_t`

When a grid is in motion, it is often necessary to account for the position of each grid point as the grid deforms. When all grid points move at the same velocity, the grid keeps its original shape. This particular case of grid motion may be recorded under the `RigidGridMotion_t` data structure described in Section 11.2. On the other hand, if the grid points have different velocity, the grid is deforming. The `ArbitraryGridMotion_t` data structure allows the CGNS file to contain information about arbitrary grid deformations. If not present, the grid is assumed to be rigid.

Note that multiple `GridCoordinates_t` nodes may be stored under a `Zone_t`. This allows the storage of the instantaneous grid locations at different time steps or iterations.

The arbitrary grid motion is recorded independently for each zone of the CGNS base. Therefore the `ArbitraryGridMotion_t` data structure is located under the zone data structure (`Zone_t`). There may be zero to several `ArbitraryGridMotion_t` nodes under a single `Zone_t` node. The multiple arbitrary grid motion definitions may be associated with different iterations or time steps in the computation. This association is recorded under the `ZoneIterativeData_t` data structure, described in Section 11.1.2.

```
ArbitraryGridMotion_t< int IndexDimension, int VertexSize[IndexDimension],
                int CellSize[IndexDimension] > :=
  {
  ArbitraryGridMotionType_t ArbitraryGridMotionType ;                    (r)

  List(DataArray_t<real, IndexDimension, DataSize[]>
     GridVelocityX GridVelocityY ... ) ;                                 (o)

  List( Descriptor_t Descriptor1 ... DescriptorN ) ;                     (o)

  GridLocation_t GridLocation ;                                          (o/d)

  Rind_t<IndexDimension> Rind ;                                          (o/d)

  DataClass_t DataClass ;                                                (o)

  DimensionalUnits_t DimensionalUnits ;                                  (o)
  }
```

Notes

1. The only required element of the `ArbitraryGridMotion_t` data structure is the `Arbitrary-GridMotionType`. Thus, even if a deforming grid application does not require the storage of grid velocity data, the `ArbitraryGridMotion_t` node must exist (with `ArbitraryGrid-MotionType = DeformingGrid`) to indicate that deformed grid points (`GridCoordinates_t`) exist for this zone.
2. `Rind` is an optional field that indicates the number of rind planes included in the grid velocity data. It only applies to structured zones.
3. The `GridLocation` specifies the location of the velocity data with respect to the grid; if absent, the data is assumed to coincide with grid vertices (i.e. `GridLocation = Vertex`).

`ArbitraryGridMotion_t` requires three structure parameters; `IndexDimension` identifies the dimensionality of the grid-size arrays, and `VertexSize` and `CellSize` are the number of 'core' vertices and cells, respectively, in each index direction. For unstructured zones, `IndexDimension` is always 1.

`ArbitraryGridMotionType_t` is an enumeration type that describes the type of arbitrary grid motion.

```
ArbitraryGridMotionType_t := Enumeration(
  Null,
  NonDeformingGrid,
  DeformingGrid,
  UserDefined ) ;
```

The `DataArray_t` nodes are used to store the components of the grid velocity vector. Table 10 lists the data-name identifiers used to record these vectors in cartesian, cylindrical, and spherical coordinate systems.

Table 10: Data-Name Identifiers for Grid Velocity

Data-Name Identifier	Description	Units
GridVelocityX	x-component of grid velocity	**L/T**
GridVelocityY	y-component of grid velocity	**L/T**
GridVelocityZ	z-component of grid velocity	**L/T**
GridVelocityR	r-component of grid velocity	**L/T**
GridVelocityTheta	θ-component of grid velocity	α/**T**
GridVelocityPhi	ϕ-component of grid velocity	α/**T**
GridVelocityXi	ξ-component of grid velocity	**L/T**
GridVelocityEta	η-component of grid velocity	**L/T**
GridVelocityZeta	ζ-component of grid velocity	**L/T**

The field `GridLocation` specifies the location of the grid velocities with respect to the grid; if absent, the grid velocities are assumed to coincide with grid vertices (i.e., `GridLocation = Vertex`). All grid velocities within a given instance of `ArbitraryGridMotion_t` must reside at the same grid location.

`Rind` is an optional field for structured zones that indicates the number of rind planes included in the data. Its purpose and function are identical to those described in Section 7.1. Note, however, that the `Rind` in this structure is independent of the `Rind` contained in `GridCoordinates_t`. They are not required to contain the same number of rind planes. Also, the location of any rind points is assumed to be consistent with the location of the 'core' data points (e.g. if `GridLocation = CellCenter`, rind points are assumed to be located at fictitious cell centers).

`DataClass` defines the default class for data contained in the `DataArray_t` entities. For dimensional grid velocities, `DimensionalUnits` may be used to describe the system of units employed. If present these two entities take precedence over the corresponding entities at higher levels of the CGNS hierarchy. The rules for determining precedence of entities of this type are discussed in Section 6.3.

Point-by-point grid velocity implies a deformation (or potentially only motion) of the grid points relative to each other. Because the original grid coordinates definition remains unchanged with the name `GridCoordinates`, any deformed coordinates must be written with a different name (e.g., `MovedGrid#1` or another used-defined name) and are pointed to using `GridCoordinatesPointers` in the data structure `ZoneIterativeData_t`, as described in Section 11.1.2.

Point-by-point grid velocity may also lead to relative motion of grid zones, or movement of grid along abutting interfaces. However, no attempt is made here to require that the `ZoneGridConnectivity_t` information be updated to be consistent with the new grid locations. Whether the `ZoneGridConnectivity_t` information refers to the original connectivity (of `GridCoordinates`) or the latest connectivity (of the moved or deformed grid) is currently left up to the user.

FUNCTION `DataSize[]`:

return value: one-dimensional `int` array of length `IndexDimension`
dependencies: `IndexDimension`, `VertexSize[]`, `CellSize[]`, `GridLocation`, `Rind`

The function `DataSize[]` is the size of the `DataArrays` containing the grid velocity components. It is identical to the function `DataSize[]` defined for `FlowSolution_t` (see Section 7.5).

11.4 Examples for Time-Dependent Flow

Example 11-A: Rigid Grid Motion

In this example, the whole mesh moves rigidly, so the only time-dependant data are the grid coordinates and flow solutions. However, since the mesh moves rigidly, the grid coordinates need not be recorded at each time step. Instead, a `RigidGridMotion_t` data structure is recorded for each step of the computation.

The number of steps and time values for each step are recorded under `BaseIterativeData_t`.

```
CGNSBase_t {
  BaseIterativeData_t {
    NumberOfSteps = N ;
    TimeValues = time1, time2, ..., timeN ;
  } ;
} ;
```

The multiple rigid grid motion and flow solution data structures are recorded under the zone. `RigidGridMotionPointers` and `FlowSolutionPointers` keep the lists of which `RigidGridMotion_t` and `FlowSolution_t` nodes correspond to each time step.

```
Zone_t Zone {

    --- Time independent data
    GridCoordinates_t GridCoordinates
    ZoneBC_t ZoneBC
    ZoneGridConnectivity_t ZoneGridConnectivity

    --- Time dependant data
    RigidGridMotion_t RigidGridMotion#1
    RigidGridMotion_t RigidGridMotion#2
    ...
    RigidGridMotion_t RigidGridmotion#N

    FlowSolution_t Solution#0
    FlowSolution_t Solution#1
    FlowSolution_t Solution#2
    ...
    FlowSolution_t Solution#N

    ZoneIterativeData_t {
      RigidGridMotionPointers = {"RigidGridMotion#1", "RigidGridMotion#2", ...,
          "RigidGridMotion#N"}
      FlowSolutionPointers = {"Solution#1", "Solution#2, ..., "Solution#N"}
    }
}
```

Note that there may be more solutions under a zone than those pointed to by `FlowSolutionPointers`. In this example, `Solution#0` could correspond to a restart solution.

Example 11-B: Deforming Grid Motion

In this example, velocity vectors are node dependant allowing for mesh deformation. In such a case, it is difficult or even impossible to recompute the mesh at each time step. Therefore the grid coordinates are recorded for each step.

Multiple `GridCoordinates_t` and `FlowSolution_t` data structures are recorded under the zone. In addition, the data structure `ArbitraryGridMotion_t` is recorded for each step. `GridCoordinatesPointers`, `FlowSolutionPointers`, and `ArbitraryGridMotionPointers_t` keep the list of which grid coordinates definition, flow solution, and arbitrary grid motion definition correspond to each time step.

```
Zone_t Zone {

    --- Time independent data
    ZoneBC_t ZoneBC
    ZoneGridConnectivity_t ZoneGridConnectivity
```

```
--- Time dependent data
List ( GridCoordinates_t GridCoordinates MovedGrid#1 MovedGrid#2 ...
        MovedGrid#N )
List ( FlowSolution_t Solution#0 Solution#1 Solution#2 ... Solution#N )
List ( ArbitraryGridMotion_t ArbitraryGridMotion#1
        ArbitraryGridMotion#2 ... ArbitraryGridMotion#N )
ZoneIterativeData_t {
  GridCoordinatesPointers = {"MovedGrid#1", "MovedGrid#2", ...,
      "MovedGrid#N"}
  FlowSolutionPointers = {"Solution#1", "Solution#2, ..., "Solution#N"}
  ArbitratyGridMotionPointers = {"ArbitraryGridMotion#1",
      "ArbitraryGridMotion#2", ..., "ArbitraryGridMotion#N"}
  }
}
```

Example 11-C: Adapted Unstructured Mesh

In this example, the mesh size varies at each remeshing, therefore new zones must be created. ZonePointers is used to keep a record of the zone definition corresponding to each recorded step. Let's assume that the solution is recorded every 50 iterations, and the grid is adapted every 100 iterations.

The number of steps, iteration values for each step, number of zones for each step, and name of these zones are recorded under BaseIterativeData_t.

```
CGNSBase_t {
  BaseIterativeData_t {
    NumberOfSteps = 4
    IterationValues = {50, 100, 150, 200}
    NumberOfZones = {1, 1, 1, 1}
    ZonePointers = {"Zone1", "Zone1", "Zone2", "Zone2"}
  }
}
```

Each zone holds 2 solutions recorded at 50 iterations apart. Therefore the ZoneIterativeData_t data structure must be included to keep track of the FlowSolutionPointers.

```
Zone_t Zone1 {

--- Constant data
GridCoordinates_t GridCoordinates
Elements_t Elements
ZoneBC_t ZoneBC
```

```
   --- Variable data
   List ( FlowSolution_t InitialSolution Solution50 Solution100 )
   ZoneIterativeData_t {
     FlowSolutionPointers = {"Solution50", "Solution100", "Null", "Null"}
   }
}

Zone_t Zone2 {

   --- Constant data
   GridCoordinates_t GridCoordinates
   Elements_t Elements
   ZoneBC_t ZoneBC

   --- Variable data
   List ( FlowSolution_t RestartSolution Solution150 Solution200 )
   ZoneIterativeData_t {
     FlowSolutionPointers = {"Null", "Null", "Solution150", "Solution200"}
   }
}
```

Notes

1. If the solution was recorded every 100 iterations instead of every 50 iterations, then each zone would have only one `FlowSolution_t` node and the data structure `ZoneIterativeData_t` would not be required.
2. Note that `FlowSolutionPointers` is always an array of size `NumberOfSteps` even if some of the steps are defined in another zone.

Example 11-D: Combination of Grid Motion and Time-Accuracy

The following is an example demonstrating the use of the rigid grid motion, arbitrary grid motion, and time-accurate data nodes in CGNS. The example is a 3-zone case. Zone 1 is rigidly rotating about the x-axis at a constant rate, with no translation. Zone 2 is a deforming zone. Zone 3 is a fixed zone. This is a time-accurate simulation with two solutions saved at times 15.5 and 31.0, corresponding to iteration numbers 1000 and 2000.

No units are given in this example, but a real case would establish them. Also, a real case would include connectivity, boundary conditions, and possibly other information as well. Each indentation represents a level down (a child) from the parent node.

```
Base (CGNSBase_t)
  SimulationType (SimulationType_t) Data=TimeAccurate
  BaseIterativeData (BaseIterativeData_t) Data=NumberOfSteps=2
    TimeValues (DataArray_t) Data=(15.5, 31.0)
    IterationValues (DataArray_t) Data=(1000, 2000)
```

```
Zone#1 (Zone_t)
  GridCoordinates (GridCoordinates_t)
    CoordinateX (DataArray_t)
    CoordinateY (DataArray_t)
  RigidGridMotion#1(RigidGridMotion_t) Data=RigidGridMotionType=ConstantRate
    OriginLocation (DataArray_t) Data=(0,0,0), (0,0,0)
    RigidRotationAngle (DataArray_t) Data=(5., 0., 0.)
  RigidGridMotion#2(RigidGridMotion_t) Data=RigidGridMotionType=ConstantRate
    OriginLocation (DataArray_t) Data=(0,0,0), (0,0,0)
    RigidRotationAngle (DataArray_t) Data=(10., 0., 0.)
  ZoneIterativeData (ZoneIterativeData_t)
    RigidGridMotionPointers (DataArray_t) Data=(RigidGridMotion#1,
                                                RigidGridMotion#2)
    FlowSolutionPointers (DataArray_t) Data=(Soln#1, Soln#2)
  Soln#1 (FlowSolution_t)
    Density (DataArray_t)
    VelocityX (DataArray_t)
  Soln#2 (FlowSolution_t)
    Density (DataArray_t)
    VelocityX (DataArray_t)
Zone#2 (Zone_t)
  GridCoordinates (GridCoordinates_t)
    CoordinateX (DataArray_t)
    CoordinateY (DataArray_t)
  MovedGrid#1 (GridCoordinates_t)
    CoordinateX (DataArray_t)
    CoordinateY (DataArray_t)
  MovedGrid#2 (GridCoordinates_t)
    CoordinateX (DataArray_t)
    CoordinateY (DataArray_t)
  ArbitraryGridMotion#1 (ArbitraryGridMotion_t)
                      Data=ArbitraryGridMotionType=DeformingGrid
  ArbitraryGridMotion#2 (ArbitraryGridMotion_t)
                      Data=ArbitraryGridMotionType=DeformingGrid
    GridVelocityX (DataArray_t)
    GridVelocityY (DataArray_t)
  ZoneIterativeData (ZoneIterativeData_t)
    ArbitraryGridMotionPointers (DataArray_t) Data=("ArbitraryGridMotion#1",
                                                "ArbitraryGridMotion#2")
    GridCoordinatesPointers (DataArray_t) Data=("MovedGrid#1",
                                              "MovedGrid#2")
    FlowSolutionPointers (DataArray_t) Data=("Soln#1", "Soln#2")
  Soln#1 (FlowSolution_t)
    Density (DataArray_t)
    VelocityX (DataArray_t)
```

```
    Soln#2 (FlowSolution_t)
      Density (DataArray_t)
      VelocityX (DataArray_t)
  Zone#3 (Zone_t)
    GridCoordinates (GridCoordinates_t)
      CoordinateX (DataArray_t)
      CoordinateY (DataArray_t)
    ZoneIterativeData (ZoneIterativeData_t)
      FlowSolutionPointers (DataArray_t) Data=("Soln#1", "Soln#2")
    Soln#1 (FlowSolution_t)
      Density (DataArray_t)
      VelocityX (DataArray_t)
    Soln#2 (FlowSolution_t)
      Density (DataArray_t)
      VelocityX (DataArray_t)
```

Notes

1. Under `BaseIterativeData_t`, one can give either `TimeValues`, or `IterationValues`, or both. In the example, both have been given.

2. The nodes `NumberOfZones` and `ZonePointers` are not required under the `BaseIterative-Data_t` node in this example because all existing zones are used for each time step.

3. Under `ArbitraryGridMotion`, the `GridVelocity` data is optional. In the example, it was put under one of the nodes but not under the other. Hence, `"ArbitraryGridMotion#1"` in the example has no children nodes, while `"ArbitraryGridMotion#2"` does.

4. The pointers under `ZoneIterativeData_t` point to names of nodes within the same zone. Thus, for example, `Soln#1` refers to the flow solution named `Soln#1` in the same zone, even though there are flow solution nodes in other zones with the same name.

5. The name `GridCoordinates` always refers to the *original* grid. Thus, when a grid is deforming, the deformed values must be put in `GridCoordinates_t` nodes of a different name. In the example, the deformed grids (for `Zone#2`) at the two times of interest were put into `"MovedGrid#1"` and `"MovedGrid#2"`.

6. Because the node `"ArbitraryGridMotion#1"` doesn't really add any information in the current example (since it was decided not to store `GridVelocity` data under it), one has the option of not including this node in the CGNS file. If it is removed, then under `Zone#2`'s `ZoneIterativeData`, the `ArbitraryGridMotionPointers` data would be replaced by:

```
Data = (Null, ArbitraryGridMotion#2)
```

12 Miscellaneous Data Structures

This section contains miscellaneous structure types for describing reference states, convergence history, discrete field data, integral or global data and families.

12.1 Reference State Structure Definition: `ReferenceState_t`

`ReferenceState_t` describes a reference state, which is a list of geometric or flow-state quantities defined at a common location or condition. Examples of typical reference states associated with CFD calculations are freestream, plenum, stagnation, inlet and exit. Note that providing a `ReferenceState` description is particularly important if items elsewhere in the CGNS database are `NormalizedByUnknownDimensional`.

```
ReferenceState_t :=
  {
  Descriptor_t ReferenceStateDescription ;                        (o)
  List( Descriptor_t Descriptor1 ... DescriptorN ) ;              (o)

  List( DataArray_t<DataType, 1, 1> DataArray1 ... DataArrayN ) ; (o)

  DataClass_t DataClass ;                                         (o)

  DimensionalUnits_t DimensionalUnits ;                           (o)
  } ;
```

Notes

1. Default names for the `Descriptor_t` and `DataArray_t` lists are as shown; users may choose other legitimate names. Legitimate names must be unique within a given instance of `ReferenceState_t` and shall not include the names `DataClass`, `DimensionalUnits`, or `ReferenceStateDescription`.

Data-name identifiers associated with `ReferenceState` are shown in Table 11.

In addition, any flowfield quantities (such as `Density`, `Pressure`, etc.) can be included in the `ReferenceState`.

The reference length L (`LengthReference`) may be necessary for `NormalizedByUnknownDimensional` databases, to define the length scale used for nondimensionalizations. It may be the same or different from the `Reynolds_Length` used to define the Reynolds number.

Because of different definitions for angle of attack and angle of yaw, these quantities are not explicitly defined in the SIDS. Instead, the user can unambigouosly denote the freestream velocity vector direction by giving `VelocityX`, `VelocityY`, and `VelocityZ` in `ReferenceState`, (with the reference state denoting the freestream).

Table 11: Data-name Identifiers for Reference State

Data-Name Identifier	Description	Units
Mach	Mach number, $M = q/c$	-
Mach_Velocity	Velocity scale, q	**L/T**
Mach_VelocitySound	Speed of sound scale, c	**L/T**
Reynolds	Reynolds number, $Re = VL_R/\nu$	-
Reynolds_Velocity	Velocity scale, V	**L/T**
Reynolds_Length	Length scale, L_R	**L**
Reynolds_ViscosityKinematic	Kinematic viscosity scale, ν	$\mathbf{L^2/T}$
LengthReference	Reference length, L	**L**

Care should be taken when defining the reference state quantities to ensure consistency. (See the discussion in Section 5.2.3.) For example, if velocity, length, and time are all defined, then the velocity stored should be length/time. If consistency is not followed, different applications could interpret the resulting data in different ways.

`DataClass` defines the default for the class of data contained in the reference state. If any reference state quantities are dimensional, `DimensionalUnits` may be used to describe the system of dimensional units employed. If present, these two entities take precedence of all corresponding entities at higher levels of the hierarchy. These precedence rules are further discussed in Section 6.3.

We recommend using the `ReferenceStateDescription` entity to document the flow conditions. The format of the documentation is currently unregulated.

12.2   Reference State Example

An example is presented in this section of a reference state entity that contains dimensional data. An additional example of a nondimensional reference state is provided in Annex B.

Example 12-A: Reference State with Dimensional Data

A freestream reference state where all data quantities are dimensional. Standard atmospheric conditions at sea level are assumed for static quantities, and all stagnation variables are obtained using the isentropic relations. The flow velocity is 200 m/s aligned with the x-axis. A consistent set of kg-m-s units are used throughout. The data class and system of units are specified at the `ReferenceState_t` level rather than attaching this information directly to the `DataArray_t` entities for each reference quantity. Data-name identifiers are provided in Annex A.

```
ReferenceState_t ReferenceState =
  {{
  Descriptor_t ReferenceStateDescription =
    {{
    Data(char, 1, 45) = "Freestream at standard atmospheric conditions" ;
```

```
    }} ;

  DataClass_t DataClass = Dimensional ;

  DimensionalUnits_t DimensionalUnits =
    {{
    MassUnits        = Kilogram ;
    LengthUnits      = Meter ;
    TimeUnits        = Second ;
    TemperatureUnits = Kelvin ;
    AngleUnits       = Radian ;
    }} ;

  DataArray_t<real, 1, 1> VelocityX =
    {{
    Data(real, 1, 1) = 200. ;
    }} ;
  DataArray_t<real, 1, 1> VelocityY             = {{ 0. }} ;
  DataArray_t<real, 1, 1> VelocityZ             = {{ 0. }} ;

  DataArray_t<real, 1, 1> Pressure              = {{ 1.0132E+05 }} ;
  DataArray_t<real, 1, 1> Density               = {{ 1.226 }} ;
  DataArray_t<real, 1, 1> Temperature           = {{ 288.15 }} ;
  DataArray_t<real, 1, 1> VelocitySound         = {{ 340. }} ;
  DataArray_t<real, 1, 1> ViscosityMolecular    = {{ 1.780E-05 }} ;

  DataArray_t<real, 1, 1> PressureStagnation    = {{ 1.2806E+05 }} ;
  DataArray_t<real, 1, 1> DensityStagnation     = {{ 1.449 }} ;
  DataArray_t<real, 1, 1> TemperatureStagnation = {{ 308.09 }} ;
  DataArray_t<real, 1, 1> VelocitySoundStagnation = {{ 351.6 }} ;

  DataArray_t<real, 1, 1> PressureDynamic       = {{ 0.2542E+05 }} ;
  }} ;
```

Note that all `DataArray_t` entities except `VelocityX` have been abbreviated.

12.3 Convergence History Structure Definition: `ConvergenceHistory_t`

Flow solver convergence history information is described by the `ConvergenceHistory_t` structure. This structure contains the number of iterations and a list of data arrays containing convergence information at each iteration.

```
  ConvergenceHistory_t :=
    {
```

```
    Descriptor_t NormDefinitions ;                                    (o)
    List( Descriptor_t Descriptor1 ... DescriptorN ) ;               (o)

    int Iterations ;                                                  (r)

    List( DataArray_t<DataType, 1, Iterations>
      DataArray1 ... DataArrayN ) ;                                   (o)

    DataClass_t DataClass ;                                           (o)

    DimensionalUnits_t DimensionalUnits ;                            (o)
    } ;
```

Notes

1. Default names for the `Descriptor_t` and `DataArray_t` lists are as shown; users may choose other legitimate names. Legitimate names must be unique within a given instance of `ConvergenceHistory_t` and shall not include the names `DataClass`, `DimensionalUnits`, or `NormDefinitions`.
2. `Iterations` is the only required field for `ConvergenceHistory_t`.

`Iterations` identifies the number of iterations for which convergence information is recorded. This value is also passed into each of the `DataArray_t` entities, defining the length of the data arrays.

`DataClass` defines the default for the class of data contained in the convergence history. If any convergence-history data is dimensional, `DimensionalUnits` may be used to describe the system of dimensional units employed. If present, these two entities take precedence over all corresponding entities at higher levels of the hierarchy. These precedence rules are further discussed in Section 6.3.

Measures used to record convergence vary greatly among current flow-solver implementations. Convergence information typically includes global forces, norms of equation residuals, and norms of solution changes. Attempts to systematically define a set of convergence measures within the CGNS project have been futile. For global parameters, such as forces and moments, Annex A lists a set of standardized data-array identifiers. For equations residuals and solution changes, no such standard list exists. It is suggested that data-array identifiers for norms of equations residuals begin with RSD, and those for solution changes begin with CHG. For example, `RSDMassRMS` could be used for the L_2-norm (RMS) of mass conservation residuals. It is also strongly recommended that `NormDefinitions` be utilized to describe the convergence information recorded in the data arrays. The format used to describe the convergence norms in `NormDefinitions` is currently unregulated.

12.4   Discrete Data Structure Definition: `DiscreteData_t`

`DiscreteData_t` provides a description of generic discrete data (i.e., data defined on a computational grid); it is identical to `FlowSolution_t` except for its type name. This structure can be used to store field data, such as fluxes or equation residuals, that is not typically considered part of the flow solution. `DiscreteData_t` contains a list for data arrays, identification of grid location, and a mechanism for identifying rind-point data included in the data arrays. All data contained within this structure must be defined at the same grid location and have the same amount of rind-point data.

```
DiscreteData_t< int IndexDimension, int VertexSize[IndexDimension],
              int CellSize[IndexDimension] > :=
  {
  List( Descriptor_t Descriptor1 ... DescriptorN ) ;                    (o)

  GridLocation_t GridLocation ;                                        (o/d)

  Rind_t<IndexDimension> Rind ;                                        (o/d)

  List( DataArray_t<DataType, IndexDimension, DataSize[]>
        DataArray1 ... DataArrayN ) ;                                   (o)

  DataClass_t DataClass ;                                              (o)

  DimensionalUnits_t DimensionalUnits ;                                (o)
  } ;
```

Notes

1. Default names for the `Descriptor_t` and `DataArray_t` lists are as shown; users may choose other legitimate names. Legitimate names must be unique within a given instance of `DiscreteData_t` and shall not include the names `DataClass`, `DimensionalUnits`, `GridLocation`, or `Rind`.
2. There are no required fields for `DiscreteData_t`. `GridLocation` has a default of `Vertex` if absent. `Rind` also has a default if absent; the the default is equivalent to having an instance of `Rind` whose `RindPlanes` array contains all zeros (see Section 4.8).
3. The structure parameter `DataType` must be consistent with the data stored in the `DataArray_t` entities (see Section 5.1).
4. For unstructured zones: rind planes are not meaningful and should not be used; `GridLocation` options are limited to `Vertex` or `CellCenter`, meaning that solution data may only be expressed at these locations; and the data arrays must follow the node ordering if `GridLocation = Vertex`, and the element ordering if `GridLocation = CellCenter`.

`DiscreteData_t` requires three structure parameters; `IndexDimension` identifies the dimensionality of the grid size arrays, and `VertexSize` and `CellSize` are the number of 'core' vertices and cells, respectively, in each index direction.

The arrays of discrete data are stored in the list of `DataArray_t` entities. The field `GridLocation` specifies the location of the data with respect to the grid; if absent, the data is assumed to coincide with grid vertices (i.e., `GridLocation = Vertex`). All data within a given instance of `DiscreteData_t` must reside at the same grid location.

Rind is an optional field that indicates the number of rind planes included in the data. Its purpose and function are identical to those described in Section 7.1. Note, however, that the Rind in this structure is independent of the Rind contained in GridCoordinates_t or FlowSolution_t. They are not required to contain the same number of rind planes. Also, the location of any rind points is assumed to be consistent with the location of the 'core' data points (e.g. if GridLocation = CellCenter, rind points are assumed to be located at fictitious cell centers).

DataClass defines the default class for data contained in the DataArray_t entities. For dimensional data, DimensionalUnits may be used to describe the system of units employed. If present these two entities take precedence over the corresponding entities at higher levels of the CGNS hierarchy. The rules for determining precedence of entities of this type are discussed in Section 6.3.

FUNCTION DataSize[]:

return value: one-dimensional int array of length IndexDimension
dependencies: IndexDimension, VertexSize[], CellSize[], GridLocation, Rind

The function DataSize[] is the size of discrete-data arrays. It is identical to the function Data-Size[] defined for FlowSolution_t (see Section 7.5).

12.5 Integral Data Structure Definition: IntegralData_t

IntegralData_t provides a description of generic global or integral data that may be associated with a particular zone or an entire database. In contrast to DiscreteData_t, integral data is not associated with any specific field location.

```
IntegralData_t :=
  {
  List( Descriptor_t Descriptor1 ... DescriptorN ) ;              (o)

  List( DataArray_t<DataType, 1, 1> DataArray1 ... DataArrayN ) ; (o)

  DataClass_t DataClass ;                                         (o)

  DimensionalUnits_t DimensionalUnits ;                           (o)
  } ;
```

Notes

1. Default names for the Descriptor_t and DataArray_t lists are as shown; users may choose other legitimate names. Legitimate names must be unique within a given instance of DiscreteData_t and shall not include the names DataClass or DimensionalUnits.
2. There are no required fields for IntegralData_t.
3. The structure parameter DataType must be consistent with the data stored in the DataArray_t entities (see Section 5.1).

`DataClass` defines the default class for data contained in the `DataArray_t` entities. For dimensional data, `DimensionalUnits` may be used to describe the system of units employed. If present these two entities take precedence over the corresponding entities at higher levels of the CGNS hierarchy. The rules for determining precedence of entities of this type are discussed in Section 6.3.

12.6   Family Data Structure Definition: `Family_t`

Geometric associations need to be set through one layer of indirection. That is, rather than setting the geometry data for each mesh entity (nodes, edges, and faces), they are associated to intermediate objects. The intermediate objects are in turn linked to nodal regions of the computational mesh. We define a CFD *family* as this intermediate object. This layer of indirection is necessary since there is rarely a 1-to-1 connection between mesh regions and geometric entities.

The `Family_t` data structure holds the CFD family data. Each mesh surface is linked to the geometric entities of the CAD databases by a name attribute. This attribute corresponds to a family of CAD geometric entities on which the mesh face is projected. Each one of these geometric entities is described in a CAD file and is not redefined within the CGNS file. A `Family_t` data structure may be included in the `CGNSBase_t` structure for each CFD family of the model.

The `Family_t` structure contains all information pertinent to a CFD family. This information includes the name attribute or family name, the boundary conditions applicable to these mesh regions, and the referencing to the CAD databases.

```
Family_t :=
  {
  List( Descriptor_t Descriptor1 ... DescriptorN ) ;                          (o)

  FamilyBC_t FamilyBC ;                                                       (o)

  List( GeometryReference_t GeometryReference1 ... GeometryReferenceN ) ; (o)

  int Ordinal ;                                                               (o)
  } ;
```

Notes

1. All data structures contained in `Family_t` are optional.
2. Default names for the `Descriptor_t` and `GeometryReference_t` lists are as shown; users may choose other legitimate names. Legitimate names must be unique at this level and must not include the names `FamilyBC` or `Ordinal`.
3. The CAD referencing data are written in the `GeometryReference_t` data structures. They identify the CAD systems and databases where the geometric definition of the family is stored.
4. The boundary condition type pertaining to a family is contained in the data structure `FamilyBC_t`. If this boundary condition type is to be used, the `BCType` specified under `BC_t` must be `FamilySpecified`.

5. For the purpose of defining zone properties, families are extended to a volume of cells. In such case, the `GeometryReference_t` structures are not used.

6. The mesh is linked to the family by attributing a family name to a BC patch or a zone in the data structure `BC_t` or `Zone_t`, respectively.

7. `Ordinal` is defined in the SIDS as a user-defined integer with no restrictions on the values that it can contain. It may be used here to attribute a number to the family.

12.7   Geometry Reference Structure Definition: `GeometryReference_t`

The standard interface data structure identifies the CAD systems used to generate the geometry, the CAD files where the geometry is stored, and the geometric entities corresponding to the family. The `GeometryReference_t` structures contain all the information necessary to associate a CFD family to the CAD databases. For each `GeometryReference_t` structure, the CAD format is recorded in `GeometryFormat`, and the CAD file in `GeometryFile`. The geometry entity or entities within this CAD file that correspond to the family are recorded under the `GeometryEntity_t` nodes.

```
GeometryReference_t :=
  {
  List( Descriptor_t Descriptor1 ... DescriptorN ) ;                    (o)

  GeometryFormat_t GeometryFormat ;                                     (o)

  GeometryFile_t GeometryFile ;                                         (o)

  List (GeometryEntity_t GeometryEntity1 ... GeometryEntityN) ;         (o/d)
  } ;
```

The `GeometryFormat` is an enumeration type that identifies the CAD system used to generate the geometry.

```
GeometryFormat_t := Enumeration(
  Null,
  NASA-IGES,
  SDRC,
  Unigraphics,
  ProEngineer,
  ICEM-CFD,
  UserDefined ) ;
```

Notes

1. All data structures contained in `GeometryReference_t` are optional. Default names for the `Descriptor_t` and `GeometryEntity_t` lists are as shown; users may choose other legitimate names. Legitimate names must be unique at this level and must not include the names `GeometryFile` or `GeometryFormat`.

2. By default, there is only one `GeometryEntity` and its name is the family name.

3. There is no limit to the number of CAD files or CAD systems referenced in a CGNS file. Different parts of the same model may be described with different CAD files of different CAD systems.

4. Other CAD geometry formats may be added to this list as needed.

12.8 Family Boundary Condition Structure Definition: `FamilyBC_t`

One of the main advantages of the concept of a layer of indirection (called a family here) is that the mesh density and the geometric entities may be modified without altering the association between nodes and intermediate objects, or between intermediate objects and geometric entities. This is very beneficial when handling boundary conditions and properties. Instead of setting boundary conditions directly on mesh entities, or on CAD entities, they may be associated to the intermediate objects. Since these intermediate objects are stable in the sense that they are not subject to mesh or geometric variations, the boundary conditions do not need to be redefined each time the model is modified. Using the concept of indirection, the boundary conditions and property settings are made independent of operations such as geometric changes, modification of mesh topology (i.e., splitting into zones), mesh refinement and coarsening, etc.

The `FamilyBC_t` data structure contains the boundary condition type. It is envisioned that it will be extended to hold both material and volume properties as well.

```
FamilyBC_t :=
  {
  BCType_t BCType;                                              (r)
  } ;
```

Annex A. Conventions for Data-Name Identifiers

Identifiers or names can be attached to `DataArray_t` entities to identify and describe the quantity being stored. To facilitate communication between different application codes, we propose to establish a set of standardized data-name identifiers with fairly precise definitions. For any identifier in this set, the associated data should be unambiguously understood. In essence, this section proposes standardized terminology for labeling CFD-related data, including grid coordinates, flow solution, turbulence model quantities, nondimensional governing parameters, boundary condition quantities, and forces and moments.

We use the convention that all standardized identifiers denote scalar quantities; this is consistent with the intended use of the `DataArray_t` structure type to describe an array of scalars. For quantities that are vectors, such as velocity, their components are listed.

Included with the lists of standard data-name identifiers, the fundamental units of dimensions associated with that quantity are provided. The following notation is used for the fundamental units: $\mathbf{M}$ is mass, $\mathbf{L}$ is length, $\mathbf{T}$ is time, Θ is temperature and α is angle. These fundamental units are directly associated with the elements of the `DimensionalExponents_t` structure. For example, a quantity that has dimensions $\mathbf{ML/T}$ corresponds to `MassExponent = +1`, `LengthExponent = +1`, and `TimeExponent = -1`.

Unless otherwise noted, all quantities in the following sections denote floating-point data types, and the appropriate `DataType` structure parameter for `DataArray_t` is `real`.

A.1 Coordinate Systems

Coordinate systems for identifying physical location are as follows:

System	3-D	2-D
Cartesian	(x, y, z)	(x, y) or (x, z) or (y, z)
Cylindrical	(r, θ, z)	(r, θ)
Spherical	(r, θ, ϕ)	
Auxiliary	(ξ, η, ζ)	(ξ, η) or (ξ, ζ) or (η, ζ)

Associated with these coordinate systems are the following unit vector conventions:

x-direction	$\hat{e}_x$	r-direction	$\hat{e}_r$	ξ-direction	$\hat{e}_\xi$
y-direction	$\hat{e}_y$	θ-direction	$\hat{e}_\theta$	η-direction	$\hat{e}_\eta$
z-direction	$\hat{e}_z$	ϕ-direction	$\hat{e}_\phi$	ζ-direction	$\hat{e}_\zeta$

Note that $\hat{e}_r$, $\hat{e}_\theta$ and $\hat{e}_\phi$ are functions of position.

We envision that one of the 'standard' coordinate systems (cartesian, cylindrical or spherical) will be used within a zone (or perhaps the entire database) to define grid coordinates and other related data. The auxiliary coordinates will be used for special quantities, including forces and moments, which may not be defined in the same coordinate system as the rest of the data. When auxiliary coordinates are used, a transformation must also be provided to uniquely define them. For example, the transform from cartesian to orthogonal auxiliary coordinates is,

$$
\begin{pmatrix} \hat{e}_\xi \\ \hat{e}_\eta \\ \hat{e}_\zeta \end{pmatrix} = \mathbf{T} \begin{pmatrix} \hat{e}_x \\ \hat{e}_y \\ \hat{e}_z \end{pmatrix},
$$

where $\mathbf{T}$ is an orthonormal matrix (2×2 in 2-D and 3×3 in 3-D).

In addition, normal and tangential coordinates are often used to define boundary conditions and data related to surfaces. The normal coordinate is identified as n with the unit vector $\hat{e}_n$.

The data-name identifiers defined for coordinate systems are listed in Table 12. All represent real `DataTypes`, except for `ElementConnectivity` and `ParentData`, which are integer.

Table 12: Data-Name Identifiers for Coordinate Systems

Data-Name Identifier	Description	Units
CoordinateX	x	L
CoordinateY	y	L
CoordinateZ	z	L
CoordinateR	r	L
CoordinateTheta	θ	α
CoordinatePhi	ϕ	α
CoordinateNormal	Coordinate in direction of $\hat{e}_n$	L
CoordinateTangential	Tangential coordinate (2-D only)	L
CoordinateXi	ξ	L
CoordinateEta	η	L
CoordinateZeta	ζ	L
CoordinateTransform	Transformation matrix ($\mathbf{T}$)	-
InterpolantsDonor	Interpolation factors	-
ElementConnectivity	Nodes making up an element	-
ParentData	Element parent identification	-

A.2 Flowfield Solution

This section describes data-name identifiers for typical Navier-Stokes solution variables. The list is obviously incomplete, but should suffice for initial implementation of the CGNS system. The variables listed in this section are dimensional or raw quantities; nondimensional parameters and coefficients based on these variables are discussed in Annex A.4.

We use fairly universal notation for state variables. Static quantities are measured with the fluid at speed: static density (ρ), static pressure (p), static temperature (T), static internal energy per unit mass (e), static enthalpy per unit mass (h), entropy (s), and static speed of sound (c). We also approximate the true entropy by the function $\tilde{s} = p/\rho^\gamma$ (this assumes an ideal gas). The velocity is $\vec{q} = u\hat{e}_x + v\hat{e}_y + w\hat{e}_z$, with magnitude $q = \sqrt{\vec{q}\cdot\vec{q}}$. Stagnation quantities are obtained by bringing the fluid isentropically to rest; these are identified by a subscript '$_0$'. The term 'total' is also used to refer to stagnation quantities.

Conservation variables are density, momentum ($\rho\vec{q} = \rho u\hat{e}_x + \rho v\hat{e}_y + \rho w\hat{e}_z$), and stagnation energy per unit volume (ρe_0).

Molecular diffusion and heat transfer introduce the molecular viscosity (μ), kinematic viscosity (ν) and thermal conductivity coefficient (k). These are obtained from the state variables through auxiliary correlations. For a perfect gas, μ and k are functions of static temperature only.

The Navier-Stokes equations involve the strain tensor ($\bar{\bar{S}}$) and the shear-stress tensor ($\bar{\bar{\tau}}$). Using indicial notation, the 3-D cartesian components of the strain tensor are,

$$\bar{\bar{S}}_{i,j} = \left(\frac{\partial u_i}{\partial x_j} + \frac{\partial u_j}{\partial x_i} \right),$$

and the stress tensor is,

$$\bar{\bar{\tau}}_{i,j} = \mu \left(\frac{\partial u_i}{\partial x_j} + \frac{\partial u_j}{\partial x_i} \right) + \lambda \frac{\partial u_k}{\partial x_k},$$

where $(x_1, x_2, x_3) = (x, y, z)$ and $(u_1, u_2, u_3) = (u, v, w)$. The bulk viscosity is usually approximated as $\lambda = -2/3\mu$.

Reynolds averaging of the Navier-Stokes equations introduce Reynolds stresses ($-\rho\overline{u'v'}$, etc.) and turbulent heat flux terms ($-\rho\overline{u'e'}$, etc.), where primed quantities are instantaneous fluctuations and the bar is an averaging operator. These quantities are obtained from auxiliary turbulence closure models. Reynolds-stress models formulate transport equations for the Reynolds stresses directly; whereas, eddy-viscosity models correlate the Reynolds stresses with the mean strain rate,

$$-\overline{u'v'} = \nu_t \left(\frac{\partial u}{\partial y} + \frac{\partial v}{\partial x} \right),$$

where ν_t is the kinematic eddy viscosity. The eddy viscosity is either correlated to mean flow quantities by algebraic models or by auxiliary transport models. An example two-equation turbulence transport model is the k-ϵ model, where transport equations are formulated for the turbulent kinetic energy ($k = \frac{1}{2}(\overline{u'u'} + \overline{v'v'} + \overline{w'w'})$) and turbulent dissipation (ϵ).

Skin friction evaluated at a surface is the dot product of the shear stress tensor with the surface normal:

$$\vec{\tau} = \bar{\bar{\tau}}\cdot\hat{n},$$

Note that skin friction is a vector.

The data-name identifiers defined for flow solution quantities are listed in Table 13.

Note that for some vector quantities, like momentum, the table only explicitly lists data-name identifiers for the x, y, and z components, and for the magnitude. It should be understood, however, that for any vector quantity with a standardized data name "Vector", the following standardized data names are also defined:

VectorX	x-component of vector
VectorY	y-component of vector
VectorZ	z-component of vector
VectorR	Radial component of vector
VectorTheta	θ-component of vector
VectorPhi	ϕ-component of vector
VectorMagnitude	Magnitude of vector
VectorNormal	Normal component of vector
VectorTangential	Tangential component of vector (2-D only)

Table 13: Data-Name Identifiers for Flow Solution Quantities

Data-Name Identifier	Description	Units
Potential	Potential: $\nabla\phi = \vec{q}$	$\mathbf{L^2/T}$
StreamFunction	Stream function (2-D): $\nabla\times\psi = \vec{q}$	$\mathbf{L^2/T}$
Density	Static density (ρ)	$\mathbf{M/L^3}$
Pressure	Static pressure (p)	$\mathbf{M/(LT^2)}$
Temperature	Static temperature (T)	Θ
EnergyInternal	Static internal energy per unit mass (e)	$\mathbf{L^2/T^2}$
Enthalpy	Static enthalpy per unit mass (h)	$\mathbf{L^2/T^2}$
Entropy	Entropy (s)	$\mathbf{ML^2/(T^2\Theta)}$
EntropyApprox	Approximate entropy ($\tilde{s} = p/\rho^\gamma$)	$\mathbf{L^{3\gamma-1}/(M^{\gamma-1}T^2)}$
DensityStagnation	Stagnation density (ρ_0)	$\mathbf{M/L^3}$
PressureStagnation	Stagnation pressure (p_0)	$\mathbf{M/(LT^2)}$
TemperatureStagnation	Stagnation temperature (T_0)	Θ
EnergyStagnation	Stagnation energy per unit mass (e_0)	$\mathbf{L^2/T^2}$
EnthalpyStagnation	Stagnation enthalpy per unit mass (h_0)	$\mathbf{L^2/T^2}$
EnergyStagnationDensity	Stagnation energy per unit volume (ρe_0)	$\mathbf{M/(LT^2)}$
VelocityX	x-component of velocity ($u = \vec{q}\cdot\hat{e}_x$)	$\mathbf{L/T}$
VelocityY	y-component of velocity ($v = \vec{q}\cdot\hat{e}_y$)	$\mathbf{L/T}$
VelocityZ	z-component of velocity ($w = \vec{q}\cdot\hat{e}_z$)	$\mathbf{L/T}$
VelocityR	Radial velocity component ($\vec{q}\cdot\hat{e}_r$)	$\mathbf{L/T}$
VelocityTheta	Velocity component in θ direction ($\vec{q}\cdot\hat{e}_\theta$)	$\mathbf{L/T}$

Continued on next page

Table 13: Data-Name Identifiers for Flow Solution Quantities (*Continued*)

Data-Name Identifier	Description	Units
VelocityPhi	Velocity component in ϕ direction $(\vec{q}\cdot\hat{e}_\phi)$	**L/T**
VelocityMagnitude	Velocity magnitude $(q = \sqrt{\vec{q}\cdot\vec{q}})$	**L/T**
VelocityNormal	Normal velocity component $(\vec{q}\cdot\hat{n})$	**L/T**
VelocityTangential	Tangential velocity component (2-D)	**L/T**
VelocitySound	Static speed of sound	**L/T**
VelocitySoundStagnation	Stagnation speed of sound	**L/T**
MomentumX	x-component of momentum (ρu)	$\mathbf{M/(L^2 T)}$
MomentumY	y-component of momentum (ρv)	$\mathbf{M/(L^2 T)}$
MomentumZ	z-component of momentum (ρw)	$\mathbf{M/(L^2 T)}$
MomentumMagnitude	Magnitude of momentum (ρq)	$\mathbf{M/(L^2 T)}$
EnergyKinetic	$(u^2 + v^2 + w^2)/2 = q^2/2$	$\mathbf{L^2/T^2}$
PressureDynamic	$\rho q^2 / 2$	$\mathbf{M/(LT^2)}$
VorticityX	$\omega_x = \partial w/\partial y - \partial v/\partial z = \vec{\omega}\cdot\hat{e}_x$	$\mathbf{T^{-1}}$
VorticityY	$\omega_y = \partial u/\partial z - \partial w/\partial x = \vec{\omega}\cdot\hat{e}_y$	$\mathbf{T^{-1}}$
VorticityZ	$\omega_z = \partial v/\partial x - \partial u/\partial y = \vec{\omega}\cdot\hat{e}_z$	$\mathbf{T^{-1}}$
VorticityMagnitude	$\omega = \sqrt{\vec{\omega}\cdot\vec{\omega}}$	$\mathbf{T^{-1}}$
SkinFrictionX	x-component of skin friction $(\vec{\tau}\cdot\hat{e}_x)$	$\mathbf{M/(LT^2)}$
SkinFrictionY	y-component of skin friction $(\vec{\tau}\cdot\hat{e}_y)$	$\mathbf{M/(LT^2)}$
SkinFrictionZ	z-component of skin friction $(\vec{\tau}\cdot\hat{e}_z)$	$\mathbf{M/(LT^2)}$
SkinFrictionMagnitude	Skin friction magnitude $(\sqrt{\vec{\tau}\cdot\vec{\tau}})$	$\mathbf{M/(LT^2)}$
VelocityAngleX	Velocity angle $(\arccos(u/q) \in [0,\,180°))$	α
VelocityAngleY	$\arccos(v/q)$	α
VelocityAngleZ	$\arccos(w/q)$	α
VelocityUnitVectorX	x-component of velocity unit vector $(\vec{q}\cdot\hat{e}_x/q)$	-
VelocityUnitVectorY	y-component of velocity unit vector $(\vec{q}\cdot\hat{e}_y/q)$	-
VelocityUnitVectorZ	z-component of velocity unit vector $(\vec{q}\cdot\hat{e}_z/q)$	-
MassFlow	Mass flow normal to a plane $(\rho\vec{q}\cdot\hat{n})$	$\mathbf{M/(L^2 T)}$
ViscosityKinematic	Kinematic viscosity $(\nu = \mu/\rho)$	$\mathbf{L^2/T}$

Continued on next page

Table 13: Data-Name Identifiers for Flow Solution Quantities (*Continued*)

Data-Name Identifier	Description	Units
ViscosityMolecular	Molecular viscosity (μ)	$\mathbf{M/(LT)}$
ViscosityEddyKinematic	Kinematic eddy viscosity (ν_t)	$\mathbf{L^2/T}$
ViscosityEddy	Eddy viscosity (μ_t)	$\mathbf{M/(LT)}$
ThermalConductivity	Thermal conductivity coefficient (k)	$\mathbf{ML/(T^3\Theta)}$
PowerLawExponent	Power-law exponent (n) in molecular viscosity or thermal conductivity model	-
SutherlandLawConstant	Sutherland's Law constant (T_s) in molecular viscosity or thermal conductivity model	Θ
TemperatureReference	Reference temperature (T_{ref}) in molecular viscosity or thermal conductivity model	Θ
ViscosityMolecularReference	Reference viscosity (μ_{ref}) in molecular viscosity model	$\mathbf{M/(LT)}$
ThermalConductivityReference	Reference thermal conductivity (k_{ref}) in thermal conductivity model	$\mathbf{ML/(T^3\Theta)}$
IdealGasConstant	Ideal gas constant $(R = c_p - c_v)$	$\mathbf{L^2/(T^2\Theta)}$
SpecificHeatPressure	Specific heat at constant pressure (c_p)	$\mathbf{L^2/(T^2\Theta)}$
SpecificHeatVolume	Specific heat at constant volume (c_v)	$\mathbf{L^2/(T^2\Theta)}$
ReynoldsStressXX	Reynolds stress $-\rho\overline{u'u'}$	$\mathbf{M/(LT^2)}$
ReynoldsStressXY	Reynolds stress $-\rho\overline{u'v'}$	$\mathbf{M/(LT^2)}$
ReynoldsStressXZ	Reynolds stress $-\rho\overline{u'w'}$	$\mathbf{M/(LT^2)}$
ReynoldsStressYY	Reynolds stress $-\rho\overline{v'v'}$	$\mathbf{M/(LT^2)}$
ReynoldsStressYZ	Reynolds stress $-\rho\overline{v'w'}$	$\mathbf{M/(LT^2)}$
ReynoldsStressZZ	Reynolds stress $-\rho\overline{w'w'}$	$\mathbf{M/(LT^2)}$
LengthReference	Reference length L	$\mathbf{L}$

A.3 Turbulence Model Solution

This section lists data-name identifiers for typical Reynolds-averaged Navier-Stokes turbulence model variables. Turbulence model solution quantities and model constants present a particularly difficult nomenclature problem—to be precise we need to identify both the variable and the model (and version) that it comes from. The list in Table 14 falls short in this respect.

Table 14: Data-Name Identifiers for Typical Turbulence Models

Data-Name Identifier	Description	Units
TurbulentDistance	Distance to nearest wall	$\mathbf{L}$
TurbulentEnergyKinetic	$k = \frac{1}{2}(\overline{u'u'} + \overline{v'v'} + \overline{w'w'})$	$\mathbf{L^2/T^2}$
TurbulentDissipation	ϵ	$\mathbf{L^2/T^3}$
TurbulentDissipationRate	$\epsilon/k = \omega$	$\mathbf{T^{-1}}$
TurbulentBBReynolds	Baldwin-Barth one-equation model R_T	-
TurbulentSANuTilde	Spalart-Allmaras one-equation model $\tilde{\nu}$	$\mathbf{L^2/T}$

A.4 Nondimensional Parameters

CFD codes are rich in nondimensional governing parameters, such as Mach number and Reynolds number, and nondimensional flowfield coefficients, such as pressure coefficient. The problem with these parameters is that their definitions and conditions that they are evaluated at can vary from code to code. Reynolds number is particularly notorious in this respect.

These parameters have posed us with a difficult dilemma. Either we impose a rigid definition for each and force all database users to abide by it, or we develop some methodology for describing the particular definition that the user is employing. The first limits applicability and flexibility, and the second adds complexity. We have opted for the second approach, but we include only enough information about the definition of each parameter to allow for conversion operations. For example, the Reynolds number includes velocity, length, and kinematic viscosity scales in its definition (i.e. $Re = VL_R/\nu$). The database description of Reynolds number includes these different scales. By providing these 'definition components', any code that reads Reynolds number from the database can transform its value to an appropriate internal definition. These 'definition components' are identified by appending a '_' to the data-name identifier of the parameter.

Definitions for nondimensional flowfield coefficients follow: the pressure coefficient is defined as,

$$c_p = \frac{p - p_{\text{ref}}}{\frac{1}{2}\rho_{\text{ref}}q_{\text{ref}}^2},$$

where $\frac{1}{2}\rho_{\text{ref}}q_{\text{ref}}^2$ is the dynamic pressure evaluated at some reference condition, and p_{ref} is some reference pressure. The skin friction coefficient is,

$$\vec{c}_f = \frac{\vec{\tau}}{\frac{1}{2}\rho_{\mathrm{ref}} q_{\mathrm{ref}}^2},$$

where $\vec{\tau}$ is the shear stress or skin friction vector. Usually, $\vec{\tau}$ is evaluated at the wall surface.

The data-name identifiers defined for nondimensional governing parameters and flowfield coefficients are listed in Table 15.

Table 15: Data-Name Identifiers for Nondimensional Parameters

Data-Name Identifier	Description	Units
Mach	Mach number: $M = q/c$	-
Mach_Velocity	Velocity scale (q)	**L/T**
Mach_VelocitySound	Speed of sound scale (c)	**L/T**
Reynolds	Reynolds number: $Re = VL_R/\nu$	-
Reynolds_Velocity	Velocity scale (V)	**L/T**
Reynolds_Length	Length scale (L_R)	**L**
Reynolds_ViscosityKinematic	Kinematic viscosity scale (ν)	$\mathbf{L^2/T}$
Prandtl	Prandtl number: $Pr = \mu c_p/k$	-
Prandtl_ThermalConductivity	Thermal conductivity scale (k)	$\mathbf{ML/(T^3\Theta)}$
Prandtl_ViscosityMolecular	Molecular viscosity scale (μ)	$\mathbf{M/(LT)}$
Prandtl_SpecificHeatPressure	Specific heat scale (c_p)	$\mathbf{L^2/(T^2\Theta)}$
PrandtlTurbulent	Turbulent Prandtl number, $\rho\nu_t c_p/k_t$	-
SpecificHeatRatio	Specific heat ratio: $\gamma = c_p/c_v$	-
SpecificHeatRatio_Pressure	Specific heat at constant pressure (c_p)	$\mathbf{L^2/(T^2\Theta)}$
SpecificHeatRatio_Volume	Specific heat at constant volume (c_v)	$\mathbf{L^2/(T^2\Theta)}$
CoefPressure	c_p	-
CoefSkinFrictionX	$\vec{c}_f \cdot \hat{e}_x$	-
CoefSkinFrictionY	$\vec{c}_f \cdot \hat{e}_y$	-
CoefSkinFrictionZ	$\vec{c}_f \cdot \hat{e}_z$	-
Coef_PressureDynamic	$\rho_{\mathrm{ref}} q_{\mathrm{ref}}^2/2$	$\mathbf{M/(LT^2)}$
Coef_PressureReference	p_{ref}	$\mathbf{M/(LT^2)}$

A.5 Characteristics and Riemann Invariants Based on 1-D Flow

Boundary condition specification for inflow/outflow or farfield boundaries often involves Riemann invariants or characteristics of the linearized inviscid flow equations. For an ideal compressible gas, these are typically defined as follows: Riemann invariants for an isentropic 1-D flow are,

$$\left[\frac{\partial}{\partial t} + (u \pm c)\frac{\partial}{\partial x} \right] \left(u \pm \frac{2}{\gamma - 1}c \right) = 0.$$

Characteristic variables for the 3-D Euler equations linearized about a constant mean flow are,

$$\left[\frac{\partial}{\partial t} + \bar{\Lambda}_n \frac{\partial}{\partial x} \right] W'_n(x, t) = 0, \qquad n = 1, 2, \ldots 5,$$

where the characteristics and corresponding characteristic variables are

Characteristic	$\bar{\Lambda}_n$	W'_n
Entropy	$\bar{u}$	$p' - \rho'/\bar{c}^2$
Vorticity	$\bar{u}$	v'
Vorticity	$\bar{u}$	w'
Acoustic	$\bar{u} \pm \bar{c}$	$p' \pm u'/(\bar{\rho}\bar{c})$

Barred quantities are evaluated at the mean flow, and primed quantities are linearized perturbations. The only non-zero mean-flow velocity component is $\bar{u}$. The data-name identifiers defined for Riemann invariants and characteristic variables are listed in Table 16.

Table 16: Data-Name Identifiers for Characteristics and Riemann Invariants

Data-Name Identifier	Description	Units
RiemannInvariantPlus	$u + 2c/(\gamma - 1)$	**L/T**
RiemannInvariantMinus	$u - 2c/(\gamma - 1)$	**L/T**
CharacteristicEntropy	$p' - \rho'/\bar{c}^2$	$\mathbf{M/(LT^2)}$
CharacteristicVorticity1	v'	**L/T**
CharacteristicVorticity2	w'	**L/T**
CharacteristicAcousticPlus	$p' + u'/(\bar{\rho}\bar{c})$	$\mathbf{M/(LT^2)}$
CharacteristicAcousticMinus	$p' - u'/(\bar{\rho}\bar{c})$	$\mathbf{M/(LT^2)}$

A.6 Forces and Moments

Conventions for data-name identifiers for forces and moments are defined in this section, even though the current SIDS provides no appropriate structures for storing these quantities. Ideally, forces and moments should be attached to geometric components or less ideally to surface grids. Neither of these have yet been incorporated into the CGNS database. We present these identifier conventions anticipating future developments of the database definition.

Given a differential force $\vec{f}$ (i.e. a force per unit area), the force integrated over a surface is,

$$\vec{F} = F_x\hat{e}_x + F_y\hat{e}_y + F_z\hat{e}_z = \int \vec{f}\,dA,$$

where $\hat{e}_x$, $\hat{e}_y$ and $\hat{e}_z$ are the unit vectors in the x, y and z directions, respectively. The moment about a point $\vec{r}_0$ integrated over a surface is,

$$\vec{M} = M_x\hat{e}_x + M_y\hat{e}_y + M_z\hat{e}_z = \int (\vec{r} - \vec{r}_0) \times \vec{f}\,dA.$$

Lift and drag components of the integrated force are,

$$L = \vec{F} \cdot \hat{L} \qquad D = \vec{F} \cdot \hat{D}$$

where $\hat{L}$ and $\hat{D}$ are the unit vectors in the positive lift and drag directions, respectively.

Lift, drag and moment are often computed in auxiliary coordinate frames (e.g. wind axes or stability axes). We introduce the convention that lift, drag and moment are computed in the (ξ, η, ζ) coordinate system. Positive drag is assumed parallel to the ξ-direction (i.e. $\hat{D} = \hat{e}_\xi$); and positive lift is assumed parallel to the η-direction (i.e. $\hat{L} = \hat{e}_\eta$). Thus, forces and moments defined in this auxiliary coordinate system are,

$$L = \vec{F} \cdot \hat{e}_\eta \qquad D = \vec{F} \cdot \hat{e}_\xi$$

$$\vec{M} = M_\xi\hat{e}_\xi + M_\eta\hat{e}_\eta + M_\zeta\hat{e}_\zeta = \int (\vec{r} - \vec{r}_0) \times \vec{f}\,dA.$$

Lift, drag and moment coefficients in 3-D are defined as,

$$C_L = \frac{L}{\frac{1}{2}\rho_{\text{ref}}q_{\text{ref}}^2 S} \qquad C_D = \frac{D}{\frac{1}{2}\rho_{\text{ref}}q_{\text{ref}}^2 S} \qquad \vec{C}_M = \frac{\vec{M}}{\frac{1}{2}\rho_{\text{ref}}q_{\text{ref}}^2 c_{\text{ref}} S_{\text{ref}}},$$

where $\frac{1}{2}\rho_{\text{ref}}q_{\text{ref}}^2$ is a reference dynamic pressure, S_{ref} is a reference area, and c_{ref} is a reference length. For a wing, S_{ref} is typically the wing area and c_{ref} is the mean aerodynamic chord. In 2-D, the sectional force coefficients are,

$$c_l = \frac{L'}{\frac{1}{2}\rho_{\text{ref}}q_{\text{ref}}^2 c_{\text{ref}}} \qquad c_d = \frac{D'}{\frac{1}{2}\rho_{\text{ref}}q_{\text{ref}}^2 c_{\text{ref}}} \qquad \vec{c}_m = \frac{\vec{M}'}{\frac{1}{2}\rho_{\text{ref}}q_{\text{ref}}^2 c_{\text{ref}}^2},$$

where the forces are integrated along a contour (e.g. an airfoil cross-section) rather than a surface.

The data-name identifiers and definitions provided for forces and moments and their associated coefficients are listed in Table 17. For coefficients, the dynamic pressure and length scales used in the normalization are provided.

Table 17: Data-Name Identifiers for Forces and Moments

Data-Name Identifier	Description	Units
ForceX	$F_x = \vec{F} \cdot \hat{e}_x$	$\mathrm{ML/T^2}$
ForceY	$F_y = \vec{F} \cdot \hat{e}_y$	$\mathrm{ML/T^2}$
ForceZ	$F_z = \vec{F} \cdot \hat{e}_z$	$\mathrm{ML/T^2}$
ForceR	$F_r = \vec{F} \cdot \hat{e}_r$	$\mathrm{ML/T^2}$
ForceTheta	$F_\theta = \vec{F} \cdot \hat{e}_\theta$	$\mathrm{ML/T^2}$
ForcePhi	$F_\phi = \vec{F} \cdot \hat{e}_\phi$	$\mathrm{ML/T^2}$
Lift	L or L'	$\mathrm{ML/T^2}$
Drag	D or D'	$\mathrm{ML/T^2}$
MomentX	$M_x = \vec{M} \cdot \hat{e}_x$	$\mathrm{ML^2/T^2}$
MomentY	$M_y = \vec{M} \cdot \hat{e}_y$	$\mathrm{ML^2/T^2}$
MomentZ	$M_z = \vec{M} \cdot \hat{e}_z$	$\mathrm{ML^2/T^2}$
MomentR	$M_r = \vec{M} \cdot \hat{e}_r$	$\mathrm{ML^2/T^2}$
MomentTheta	$M_\theta = \vec{M} \cdot \hat{e}_\theta$	$\mathrm{ML^2/T^2}$
MomentPhi	$M_\phi = \vec{M} \cdot \hat{e}_\phi$	$\mathrm{ML^2/T^2}$
MomentXi	$M_\xi = \vec{M} \cdot \hat{e}_\xi$	$\mathrm{ML^2/T^2}$
MomentEta	$M_\eta = \vec{M} \cdot \hat{e}_\eta$	$\mathrm{ML^2/T^2}$
MomentZeta	$M_\zeta = \vec{M} \cdot \hat{e}_\zeta$	$\mathrm{ML^2/T^2}$
Moment_CenterX	$x_0 = \vec{r}_0 \cdot \hat{e}_x$	L
Moment_CenterY	$y_0 = \vec{r}_0 \cdot \hat{e}_y$	L
Moment_CenterZ	$z_0 = \vec{r}_0 \cdot \hat{e}_z$	L
CoefLift	C_L or c_l	-
CoefDrag	C_D or c_d	-
CoefMomentX	$\vec{C}_M \cdot \hat{e}_x$ or $\vec{c}_m \cdot \hat{e}_x$	-
CoefMomentY	$\vec{C}_M \cdot \hat{e}_y$ or $\vec{c}_m \cdot \hat{e}_y$	-
CoefMomentZ	$\vec{C}_M \cdot \hat{e}_z$ or $\vec{c}_m \cdot \hat{e}_z$	-
CoefMomentR	$\vec{C}_M \cdot \hat{e}_r$ or $\vec{c}_m \cdot \hat{e}_r$	-
CoefMomentTheta	$\vec{C}_M \cdot \hat{e}_\theta$ or $\vec{c}_m \cdot \hat{e}_\theta$	-
CoefMomentPhi	$\vec{C}_M \cdot \hat{e}_\phi$ or $\vec{c}_m \cdot \hat{e}_\phi$	-
CoefMomentXi	$\vec{C}_M \cdot \hat{e}_\xi$ or $\vec{c}_m \cdot \hat{e}_\xi$	-
CoefMomentEta	$\vec{C}_M \cdot \hat{e}_\eta$ or $\vec{c}_m \cdot \hat{e}_\eta$	-
CoefMomentZeta	$\vec{C}_M \cdot \hat{e}_\zeta$ or $\vec{c}_m \cdot \hat{e}_\zeta$	-
Coef_PressureDynamic	$1/2 \rho_{\mathrm{ref}} q_{\mathrm{ref}}^2$	$\mathrm{M/(LT^2)}$
Coef_Area	S_{ref}	$\mathrm{L^2}$
Coef_Length	c_{ref}	L

A.7 Time-Dependent Flow

Data-name identifiers related to time-dependent flow include those associated with the storage of grid coordinates and flow solutions as a function of time level or iteration. Also included are identifiers for storing information defining both rigid and arbitrary (i.e., deforming) grid motion.

Table 18: Data-Name Identifiers for Time-Dependent Flow

Data-Name Identifier	Data Type	Description	Units
TimeValues	real	Time values	T
IterationValues	int	Iteration values	-
NumberOfZones	int	Number of zones used for each recorded step	-
NumberOfFamilies	int	Number of families used for each recorded step	-
ZonePointers	char	Names of zones used for each recorded step	-
FamilyPointers	char	Names of families used for each recorded step	-
RigidGridMotionPointers	char	Names of RigidGridMotion structures used for each recorded step for a zone	-
ArbitraryGridMotionPointers	char	Names of ArbitraryGridMotion structures used for each recorded step for a zone	-
GridCoordinatesPointers	char	Names of GridCoordinates structures used for each recorded step for a zone	-
FlowSolutionsPointers	char	Names of FlowSolutions structures used for each recorded step for a zone	-
OriginLocation	real	Physical coordinates of the origin before and after a rigid grid motion	L
RigidRotationAngle	real	Rotation angles about each axis of the translated coordinate system for rigid grid motion	α
RigidVelocity	real	Grid velocity vector of the origin translation for rigid grid motion	L/T
RigidRotationRate	real	Rotation rate vector about the axis of the translated coordinate system for rigid grid motion	α/T
GridVelocityX	real	x-component of grid velocity	L/T
GridVelocityY	real	y-component of grid velocity	L/T
GridVelocityZ	real	z-component of grid velocity	L/T

Continued on next page

Table 18: Data-Name Identifiers for Time-Dependent Flow (*Continued*)

Data-Name Identifier	Data Type	Description	Units
GridVelocityR	real	r-component of grid velocity	L/T
GridVelocityTheta	real	θ-component of grid velocity	α/T
GridVelocityPhi	real	ϕ-component of grid velocity	α/T
GridVelocityXi	real	ξ-component of grid velocity	L/T
GridVelocityEta	real	η-component of grid velocity	L/T
GridVelocityZeta	real	ζ-component of grid velocity	L/T

Annex B. Structured Two-Zone Flat Plate Example

This section describes a complete database for a sample test case. The test case is compressible turbulent flow past a flat plat at zero incidence. The domain is divided into two zones as shown in Figure 6. The interface between the two zones is 1-to-1.

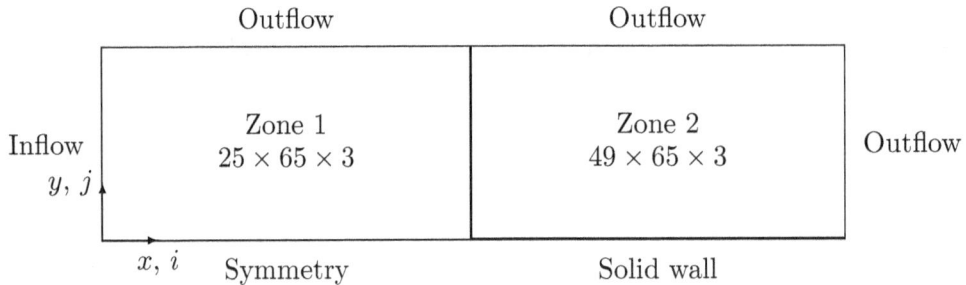

Figure 6: Two-Zone Flat Plate Test Case

The database description includes the following:

- range of indices within each zone

- grid coordinates of vertices

- flowfield solution at cell centers including a row of ghost-cells along each boundary; the flowfield includes the conservation variables and a turbulent transport variable

- multizone interface connectivity information

- boundary condition information

- reference state

- description of the compressible Navier-Stokes equations including one-equation turbulence model

Each of these items is described in separate sections to make the information more readable. The same database is presented in each section, but only that information needed for the particular focus is included. The overall layout of the database is presented in Annex B.1.

All data for this test case is nondimensional and is normalized consistently by the following (dimensional) quantities: plate length L, freestream static density ρ_∞, freestream static speed of sound c_∞, and freestream static temperature T_∞. The fact that the database is completely nondimensional is reflected in the value of the globally set data class.

B.1 Overall Layout

This section describes the overall layout of the database. Included are the cell dimension and physical dimension of the grid, the globally set data class, the global reference state and flow-equations description, and data pertaining to each zone. Each zone contains the grid size, grid coordinates, flow solution, multizone interfaces and boundary conditions. All entities given by {{*}} are expanded in subsequent sections. Note that because this example contains structured zones, IndexDimension = CellDimension = 3 in each zone.

```
CGNSBase_t TwoZoneCase =
  {{
  int CellDimension = 3 ;
  int PhysicalDimension = 3 ;

  DataClass_t DataClass = NormalizedByUnknownDimensional ;

  ReferenceState_t ReferenceState = {{*}} ;

  FlowEquationSet_t<3> FlowEquationSet = {{*}} ;

  !  CellDimension = 3, PhysicalDimension = 3
  Zone_t<3,3> Zone1 =
    {{
    int VertexSize = [25,65,3] ;
    int CellSize   = [24,64,2] ;
    int VertexSizeBoundary = [0,0,0];

    ZoneType_t ZoneType = Structured;

    !  IndexDimension = 3
    GridCoordinates_t<3,VertexSize> GridCoordinates = {{*}} ;

    FlowSolution_t<3,VertexSize,CellSize> FlowSolution = {{*}} ;

    ZoneGridConnectivity_t<3,3> ZoneGridConnectivity = {{*}} ;

    ZoneBC_t<3,3> ZoneBC = {{*}} ;
    }} ;           ! end Zone1

  !  CellDimension = 3, PhysicalDimension = 3
  Zone_t<3,3> Zone2 =
    {{
    int VertexSize = [49,65,3] ;
    int CellSize   = [48,64,2] ;
    int VertexSizeBoundary = [0,0,0];
```

```
      ZoneType_t ZoneType = Structured;

      !  IndexDimension = 3
      GridCoordinates_t<3,VertexSize> GridCoordinates = {{*}} ;

      FlowSolution_t<3,VertexSize,CellSize> FlowSolution = {{*}} ;

      ZoneGridConnectivity_t<3,3> ZoneGridConnectivity = {{*}} ;

      ZoneBC_t<3,3> ZoneBC = {{*}} ;
    }} ;           ! end Zone2
  }} ;             ! end TwoZoneCase
```

B.2 Grid Coordinates

This section describes the grid-coordinate entities for each zone. Since the coordinates are all nondimensional, the individual **DataArray_t** entities do not include a data-class qualifier; instead, this information is derived from the globally set data class. The grid-coordinate entities for zone 2 are abbreviated.

```
CGNSBase_t TwoZoneCase =
  {{
  int CellDimension = 3 ;
  int PhysicalDimension = 3 ;

  DataClass_t DataClass = NormalizedByUnknownDimensional ;

  !  CellDimension = 3, PhysicalDimension = 3
  Zone_t<3,3> Zone1 =
    {{
    int VertexSize = [25,65,3] ;
    int CellSize   = [24,64,2] ;
    int VertexSizeBoundary = [0,0,0];

    ZoneType_t ZoneType = Structured;

    !  IndexDimension = 3
    !  VertexSize = [25,65,3]
    GridCoordinates_t<3, [25,65,3]> GridCoordinates =
      {{
      DataArray_t<real, 3, [25,65,3]> CoordinateX =
        {{
        Data(real, 3, [25,65,3]) = (((x(i,j,k), i=1,25), j=1,65), k=1,3) ;
```

```
      }} ;

   DataArray_t<real, 3, [25,65,3]> CoordinateY =
     {{
     Data(real, 3, [25,65,3]) = (((y(i,j,k), i=1,25), j=1,65), k=1,3) ;
     }} ;

   DataArray_t<real, 3, [25,65,3]> CoordinateZ =
     {{
     Data(real, 3, [25,65,3]) = (((z(i,j,k), i=1,25), j=1,65), k=1,3) ;
     }} ;
   }} ;          ! end Zone1/GridCoordinates
 }} ;            ! end Zone1

!  CellDimension = 3, PhysicalDimension = 3
Zone_t<3,3> Zone2 =
  {{
  int VertexSize = [49,65,3] ;
  int CellSize    = [48,64,2] ;
  int VertexSizeBoundary = [0,0,0];

  ZoneType_t ZoneType = Structured;

  !  IndexDimension = 3
  !  VertexSize = [49,65,3]
  GridCoordinates_t<3, [49,65,3]> GridCoordinates =
    {{
    DataArray_t<real, 3, [49,65,3]> CoordinateX = {{*}} ;
    DataArray_t<real, 3, [49,65,3]> CoordinateY = {{*}} ;
    DataArray_t<real, 3, [49,65,3]> CoordinateZ = {{*}} ;
    }} ;          ! end Zone2/GridCoordinates
  }} ;            ! end Zone2
 }} ;             ! end TwoZoneCase
```

B.3 Flowfield Solution

This section provides a description of the flowfield solution including the conservation variables and the Spalart-Allmaras turbulent-transport quantity ($\tilde{\nu}$). The flowfield solution is given at cell centers with a single row of ghost-cell values along each boundary.

As with the case for grid coordinates, the flow solution is nondimensional, and this fact is derived from the globally set data class. The normalizations for each flow variable are,

$$\rho'_{ijk} = \frac{\rho_{ijk}}{\rho_\infty}, \qquad (\rho u)'_{ijk} = \frac{(\rho u)_{ijk}}{\rho_\infty c_\infty}, \qquad (\rho e_0)'_{ijk} = \frac{(\rho e_0)_{ijk}}{\rho_\infty c_\infty^2}, \qquad \tilde{\nu}'_{ijk} = \frac{\tilde{\nu}_{ijk}}{c_\infty L},$$

where primed quantities are nondimensional and all others are dimensional.

Only the `Density` entity for zone 1 is fully described in the following. The momentum, energy and turbulence solution are abbreviated. The entire flow-solution data for zone 2 is also abbreviated.

```
CGNSBase_t TwoZoneCase =
  {{
  int CellDimension = 3 ;
  int PhysicalDimension = 3 ;

  DataClass_t DataClass = NormalizedByUnknownDimensional ;

  !  CellDimension = 3, PhysicalDimension = 3
  Zone_t<3,3> Zone1 =
    {{
    int VertexSize = [25,65,3] ;
    int CellSize   = [24,64,2] ;
    int VertexSizeBoundary = [0,0,0];

    ZoneType_t ZoneType = Structured;

    !  IndexDimension = 3
    !  VertexSize = [25,65,3]
    !  CellSize   = [24,64,2]
    FlowSolution_t<3, [25,65,3], [24,64,2]> FlowSolution =
      {{
      GridLocation_t GridLocation = CellCenter ;

      !  IndexDimension = 3
      Rind_t<3> Rind =
        {{
        int[6] RindPlanes = [1,1,1,1,1,1] ;
        }} ;

      !  IndexDimension = 3
      !  DataSize = CellSize + [2,2,2] = [26,66,4]
      DataArray_t<real, 3, [26,66,4]> Density =
        {{
        Data(real, 3, [26,66,4]) = (((rho(i,j,k), i=0,25), j=0,65), k=0,3) ;
        }} ;

      DataArray_t<real, 3, [26,66,4]> MomentumX = {{*}} ;
      DataArray_t<real, 3, [26,66,4]> MomentumY = {{*}} ;
      DataArray_t<real, 3, [26,66,4]> MomentumZ = {{*}} ;
      DataArray_t<real, 3, [26,66,4]> EnergyStagnationDensity = {{*}} ;
      DataArray_t<real, 3, [26,66,4]> TurbulentSANutilde = {{*}} ;
```

```
   }} ;          ! end Zone1/FlowSolution
  }} ;           ! end Zone1

Zone_t<3,3> Zone2 = {{*}} ;
}} ;             ! end TwoZoneCase
```

B.4 Interface Connectivity

This section describes the interface connectivity between zones 1 and 2; it also includes the k-plane periodicity for each zone (which is essentially an interface connectivity of a zone onto itself). Each interface entity is repeated with the receiver and donor-zone roles reversed; this includes the periodic k-plane interfaces. Since each interface is a complete zone face, the GridConnectivity1to1_t entities are named after the face.

Because of the orientation of the zones, the index transformation matrices (Transform) for all interfaces are diagonal. This means that each matrix is its own inverse, and the value of Transform is the same for every pair of interface entities.

```
CGNSBase_t TwoZoneCase =
  {{
  int CellDimension = 3 ;
  int PhysicalDimension = 3 ;


  !  -----  ZONE 1 Interfaces  ------

  !  CellDimension = 3, PhysicalDimension = 3
  Zone_t<3,3> Zone1 =
    {{
    int VertexSize = [25,65,3] ;
    int CellSize   = [24,64,2] ;
    int VertexSizeBoundary = [0,0,0];

    ZoneType_t ZoneType = Structured;

    !  IndexDimension = 3, CellDimension = 3
    ZoneGridConnectivity_t<3,3> ZoneGridConnectivity =
      {{

      !  IndexDimension = 3
      GridConnectivity1to1_t<3> IMax =                    ! ZONE 1 IMax
        {{
        int[3] Transform = [1,2,3] ;
        IndexRange_t<3> PointRange =
```

```
      {{
      int[3] Begin = [25,1 ,1] ;
      int[3] End   = [25,65,3] ;
      }} ;
    IndexRange_t<3> PointRangeDonor =
      {{
      int[3] Begin = [1,1 ,1] ;
      int[3] End   = [1,65,3] ;
      }} ;
    Identifier(Zone_t) ZoneDonorName = Zone2 ;
    }} ;

  GridConnectivity1to1_t<3> KMin =                    ! ZONE 1 KMin
    {{
    int[3] Transform = [1,2,-3] ;
    IndexRange_t<3> PointRange =
      {{
      int[3] Begin = [1 ,1 ,1] ;
      int[3] End   = [25,65,1] ;
      }} ;
    IndexRange_t<3> PointRangeDonor =
      {{
      int[3] Begin = [1 ,1 ,3] ;
      int[3] End   = [25,65,3] ;
      }} ;
    Identifier(Zone_t) ZoneDonorName = Zone1 ;
    }} ;

  GridConnectivity1to1_t<3> KMax =                    ! ZONE 1 KMax
    {{
    int[3] Transform = [1,2,-3] ;
    IndexRange_t<3> PointRange =
      {{
      int[3] Begin = [1 ,1 ,3] ;
      int[3] End   = [25,65,3] ;
      }} ;
    IndexRange_t<3> PointRangeDonor =
      {{
      int[3] Begin = [1 ,1 ,1] ;
      int[3] End   = [25,65,1] ;
      }} ;
    Identifier(Zone_t) ZoneDonorName = Zone1 ;
    }} ;
  }} ;        ! end Zone1/ZoneGridConnectivity
}} ;          ! end Zone1
```

```
!  -----   ZONE 2 Interfaces  ------

!  CellDimension = 3, PhysicalDimension = 3
 Zone_t<3,3> Zone2 =
  {{
  int VertexSize = [49,65,3] ;
  int CellSize   = [48,64,2] ;
  int VertexSizeBoundary = [0,0,0];

  ZoneType_t ZoneType = Structured;

  !  IndexDimension = 3, CellDimension = 3
  ZoneGridConnectivity_t<3,3> ZoneGridConnectivity =
    {{

   !  IndexDimension = 3
    GridConnectivity1to1_t<3> IMin =                   ! ZONE 2 IMin
      {{
      int[3] Transform = [1,2,3] ;
      IndexRange_t<3> PointRange =
        {{
        int[3] Begin = [1,1 ,1] ;
        int[3] End   = [1,65,3] ;
        }} ;
      IndexRange_t<3> PointRangeDonor =
        {{
        int[3] Begin = [25,1 ,1] ;
        int[3] End   = [25,65,3] ;
        }} ;
      Identifier(Zone_t) ZoneDonorName = Zone1 ;
      }} ;

    GridConnectivity1to1_t<3> KMin =                   ! ZONE 2 KMin
      {{
      int[3] Transform = [1,2,-3] ;
      IndexRange_t<3> PointRange =
        {{
        int[3] Begin = [1 ,1 ,1] ;
        int[3] End   = [49,65,1] ;
        }} ;
      IndexRange_t<3> PointRangeDonor =
        {{
        int[3] Begin = [1 ,1 ,3] ;
```

```
            int[3] End   = [49,65,3] ;
            }} ;
         Identifier(Zone_t) ZoneDonorName = Zone2 ;
         }} ;

      GridConnectivity1to1_t<3> KMax =                    ! ZONE 2 KMax
         {{
         int[3] Transform = [1,2,-3] ;
         IndexRange_t<3> PointRange =
            {{
            int[3] Begin = [1 ,1 ,3] ;
            int[3] End   = [49,65,3] ;
            }} ;
         IndexRange_t<3> PointRangeDonor =
            {{
            int[3] Begin = [1 ,1 ,1] ;
            int[3] End   = [49,65,1] ;
            }} ;
         Identifier(Zone_t) ZoneDonorName = Zone2 ;
         }} ;
      }} ;        ! end Zone2/ZoneGridConnectivity
   }} ;          ! end Zone2
 }} ;            ! end TwoZoneCase
```

B.5 Boundary Conditions

Boundary conditions for the flat plate case are described in this section. The minimal information necessary is included in each boundary condition; this includes the boundary-condition type and BC-patch specification. The lone exception is the viscous wall, which is isothermal and has an imposed temperature profile (given by the array `temperatureprofile()`). For all other boundary conditions a flow solver is free to impose appropriate BC-data since none is provided in the following. The imposed BC-data for all cases should be evaluated at the globally set reference state, since no other reference states have been specified.

No boundary condition descriptions are provided for the multizone interface or for the k-plane periodicity in each zone. All relevant information is provided for these interfaces in the `GridConnectivity1to1_t` entities of the previous section.

The practice of naming `BC_t` entities after the face is followed.

```
CGNSBase_t TwoZoneCase =
  {{
  int CellDimension = 3 ;
  int PhysicalDimension = 3 ;
```

```
DataClass_t DataClass = NormalizedByUnknownDimensional ;

! -----   ZONE 1 BC's  ------

!  CellDimension = 3, PhysicalDimension = 3
Zone_t<3,3> Zone1 =
  {{
  int VertexSize = [25,65,3] ;
  int CellSize   = [24,64,2] ;
  int VertexSizeBoundary = [0,0,0];

  ZoneType_t ZoneType = Structured;

  !  IndexDimension = 3, PhysicalDimension = 3
  ZoneBC_t<3,3> ZoneBC =
    {{

    !  IndexDimension = 3, PhysicalDimension = 3
    BC_t<3,3> IMin =                                   !  ZONE 1 IMin
      {{
      BCType_t BCType = BCInflowSubsonic ;
      IndexRange_t<3> PointRange =
        {{
        int[3] Begin = [1,1 ,1] ;
        int[3] End   = [1,65,3] ;
        }} ;
      }} ;

    BC_t<3,3> JMin =                                   !  ZONE 1 JMin
      {{
      BCType_t BCType = BCSymmetryPlane ;
      IndexRange_t<3> PointRange =
        {{
        int[3] Begin = [1 ,1,1] ;
        int[3] End   = [25,1,3] ;
        }} ;
      }} ;

    BC_t<3,3> JMax =                                   !  ZONE 1 JMax
      {{
      BCType_t BCType = BCOutFlowSubsonic ;
      IndexRange_t<3> PointRange =
        {{
        int[3] Begin = [1 ,65,1] ;
        int[3] End   = [25,65,3] ;
```

```
        }} ;
      }} ;
    }} ;          ! end Zone1/ZoneBC
  }} ;            ! end Zone1


! -----   ZONE 2 BC's  ------

!  CellDimension = 3, PhysicalDimension = 3
Zone_t<3,3> Zone2 =
  {{
  int VertexSize = [49,65,3] ;
  int CellSize   = [48,64,2] ;
  int VertexSizeBoundary = [0,0,0];


  ZoneType_t ZoneType = Structured;

  !  IndexDimension = 3, PhysicalDimension = 3
  ZoneBC_t<3,3> ZoneBC =
    {{

    !  IndexDimension = 3, PhysicalDimension = 3
    BC_t<3,3> IMax =                              ! ZONE 2 IMax
      {{
      BCType_t BCType = BCOutflowSubsonic ;
      IndexRange_t<3> PointRange =
        {{
        int[3] Begin = [49,1 ,1] ;
        int[3] End   = [49,65,3] ;
        }} ;
      }} ;    ! end Zone2/ZoneBC/IMax

    BC_t<3,3> JMin =                              ! ZONE 2 JMin
      {{
      BCType_t BCType = BCWallViscous ;
      IndexRange_t<3> PointRange =
        {{
        int[3] Begin = [1 ,1,1] ;
        int[3] End   = [49,1,3] ;
        }} ;

      !  ListLength = 49*3 = 147
      BCDataSet<147> BCDataSet =
        {{
        BCTypeSimple_t BCTypeSimple = BCWallViscousIsothermal ;
```

```
         !  Data array length = ListLength = 147
      BCData_t<147> DirichletData =
        {{
        DataArray_t<real, 1, 147> Temperature =
          {{
          Data(real, 1, 147) = (temperatureprofile(n), n=1,147) ;
          }} ;
        }} ;
      }} ;
    }} ;     ! end Zone2/ZoneBC/JMin

  BC_t<3,3> JMax =                                    !  ZONE 2 JMax
    {{
    BCType_t BCType = BCOutFlowSubsonic ;
    IndexRange_t<3> PointRange =
      {{
      int[3] Begin = [1 ,65,1] ;
      int[3] End   = [49,65,3] ;
      }} ;
    }} ;     ! end Zone2/ZoneBC/JMax

    }} ;      ! end Zone2/ZoneBC
  }} ;        ! end Zone2
}} ;          ! end TwoZoneCase
```

B.6 Global Reference State

This section provides a description of the freestream reference state. As previously stated, all data is nondimensional including all reference state quantities. The dimensional plate length L and freestream scales ρ_∞, c_∞ and T_∞ are used for normalization.

The freestream Mach number is 0.5 and the Reynolds number is 10^6 based on freestream velocity and kinematic viscosity and the plate length. These are the only nondimensional parameters included in the reference state. The defining scales for each parameter are also included; these defining scales are nondimensional.

Using consistent normalization, the following nondimensional freestream quantities are defined:

$$\rho'_\infty = 1 \qquad (\rho_0)'_\infty = \rho'_\infty \Gamma^{1/(\gamma-1)} \qquad L' = 1$$
$$c'_\infty = 1 \qquad (c_0)'_\infty = c'_\infty \Gamma^{1/2} \qquad u'_\infty = M_\infty = 0.5$$
$$T'_\infty = 1 \qquad (T_0)'_\infty = T'_\infty \Gamma \qquad v'_\infty = 0$$
$$p'_\infty = 1/\gamma \qquad (p_0)'_\infty = p'_\infty \Gamma^{\gamma/(\gamma-1)} \qquad w'_\infty = 0$$
$$e'_\infty = 1/\gamma(\gamma-1) \qquad (e_0)'_\infty = e'_\infty \Gamma \qquad \nu'_\infty = u'_\infty L'/Re = 5\times10^{-7}$$
$$h'_\infty = 1/(\gamma-1) \qquad (h_0)'_\infty = h'_\infty \Gamma \qquad \tilde{s}'_\infty = p'_\infty/(\rho'_\infty)^\gamma = 1/\gamma$$

where $\Gamma \equiv 1 + \frac{\gamma-1}{2}M_\infty^2$ based on $M_\infty = 0.5$ and $\gamma = 1.4$.

Except for the nondimensional parameters Mach number and Reynolds number, all `DataArray_t` entities are abbreviated.

```
CGNSBase_t TwoZoneCase =
  {{

  DataClass_t DataClass = NormalizedByUnknownDimensional ;

  ReferenceState_t ReferenceState =
    {{
    Descriptor_t ReferenceStateDescription =
      {{
      Data(char, 1, 10) = "Freestream" ;
      }} ;

    DataArray_t<real, 1, 1> Mach =
      {{
      Data(real, 1, 1) = 0.5 ;
      DataClass_t DataClass = NondimensionalParameter ;
      }} ;
    DataArray_t<real, 1, 1> Mach_Velocity      = {{ 0.5 }} ;
    DataArray_t<real, 1, 1> Mach_VelocitySound = {{ 1 }} ;

    DataArray_t<real, 1, 1> Reynolds =
      {{
      Data(real, 1, 1) = 1.0e+06 ;
      DataClass_t DataClass = NondimensionalParameter ;
      }} ;
    DataArray_t<real, 1, 1> Reynolds_Velocity          = {{ 0.5 }} ;
    DataArray_t<real, 1, 1> Reynolds_Length            = {{ 1. }} ;
    DataArray_t<real, 1, 1> Reynolds_ViscosityKinematic = {{ 5.0E-07 }} ;

    DataArray_t<real, 1, 1> Density          = {{ 1. }} ;
    DataArray_t<real, 1, 1> LengthReference  = {{ 1. }} ;
    DataArray_t<real, 1, 1> VelocitySound    = {{ 1. }} ;
    DataArray_t<real, 1, 1> VelocityX        = {{ 0.5 }} ;
    DataArray_t<real, 1, 1> VelocityY        = {{ 0 }};
    DataArray_t<real, 1, 1> VelocityZ        = {{ 0 }} ;
    DataArray_t<real, 1, 1> Pressure         = {{ 0.714286 }} ;
    DataArray_t<real, 1, 1> Temperature      = {{ 1. }} ;
    DataArray_t<real, 1, 1> EnergyInternal   = {{ 1.785714 }} ;
    DataArray_t<real, 1, 1> Enthalpy         = {{ 2.5 }} ;
    DataArray_t<real, 1, 1> EntropyApprox    = {{ 0.714286 }} ;
```

```
    DataArray_t<real, 1, 1> DensityStagnation       = {{ 1.129726 }} ;
    DataArray_t<real, 1, 1> PressureStagnation      = {{ 0.847295 }} ;
    DataArray_t<real, 1, 1> EnergyStagnation        = {{ 1.875 }} ;
    DataArray_t<real, 1, 1> EnthalpyStagnation      = {{ 2.625 }} ;
    DataArray_t<real, 1, 1> TemperatureStagnation   = {{ 1.05 }} ;
    DataArray_t<real, 1, 1> VelocitySoundStagnation = {{ 1.024695 }} ;

    DataArray_t<real, 1, 1> ViscosityKinematic      = {{ 5.0E-07 }} ;
    }} ;
}} ;            ! end TwoZoneCase
```

B.7 Equation Description

This section provides a description of the flow equations used to solve the problem. The flow equation set is turbulent, compressible 3-D Navier-Stokes with the Spalart-Allmaras (S-A) one-equation turbulence model. The thin-layer Navier-Stokes diffusion terms are modeled; only diffusion in the j-coordinate direction is included.

A perfect gas assumption is made with $\gamma = 1.4$; based on the normalization used in this database, the nondimensional scales defining γ are $(c_p)' = 1/(\gamma-1)$ and $(c_v)' = 1/\gamma(\gamma-1)$. The molecular viscosity is obtained from Sutherland's Law. In order to nondimensionalize the viscosity formula, standard atmospheric conditions are assumed (i.e. $T_\infty = 288.15$ K). A constant Prandtl number assumption is made for the thermal conductivity coefficient; $Pr = 0.72$. The defining scales of Pr are evaluated at freestream conditions; the nondimensional thermal conductivity is $k'_\infty = \mu'_\infty (c_p)'/Pr$.

The Navier-Stokes equations are closed with an eddy viscosity assumption using the S-A model. A turbulent Prandtl number of $Pr_t = 0.9$ is prescribed. All parameters not provided are defaulted.

Except for the nondimensional parameters γ and Pr, all `DataArray_t` entities are abbreviated.

```
CGNSBase_t TwoZoneCase =
  {{
  int CellDimension = 3 ;
  int PhysicalDimension = 3 ;

  DataClass_t DataClass = NormalizedByUnknownDimensional ;

  !  CellDimension = 3
  FlowEquationSet_t<3> FlowEquationSet =
    {{
    int EquationDimension = 3

    !  CellDimension = 3 ;
    GoverningEquations_t<3> GoverningEquations =
      {{
      GoverningEquationsType_t GoverningEquationsType = NSTurbulent ;
```

```
    int[6] DiffusionModel = [0,1,0,0,0,0] ;
    }} ;

GasModel_t GasModel =
  {{
  GasModelType_t GasModelType = Ideal ;

  DataArray_t<real, 1, 1> SpecificHeatRatio =
    {{
    Data(real, 1, 1) = 1.4 ;
    DataClass_t DataClass = NondimensionalParameter ;
    }} ;
  DataArray_t<real, 1, 1> SpecificHeatRatio_Pressure = {{ 2.5 }} ;
  DataArray_t<real, 1, 1> SpecificHeatRatio_Volume   = {{ 1.785714 }} ;
  }} ;

ViscosityModel_t ViscosityModel =
  {{
  ViscosityModelType_t ViscosityModelType = SutherLandLaw ;

  DataArray_t<real, 1, 1> SutherLandLawConstant       = {{ 0.38383 }} ;
  DataArray_t<real, 1, 1> TemperatureReference        = {{ 1.05491 }} ;
  DataArray_t<real, 1, 1> ViscosityMolecularReference = {{ 5.0E-07 }} ;
  }} ;

ThermalConductivityModel_t ThermalConductivityModel =
  {{
  ThermalConductivityModelType_t ThermalConductivityModelType =
    ConstantPrandtl ;

  DataArray_t<real, 1, 1> Prandtl =
    {{
    Data(real, 1, 1) = 0.72 ;
    DataClass_t DataClass = NondimensionalParameter ;
    }} ;
  DataArray_t<real, 1, 1> Prandtl_ThermalConductivity = {{ 1.73611E-0.6 }} ;
  DataArray_t<real, 1, 1> Prandtl_ViscosityMolecular  = {{ 5.0E-0.7 }} ;
  DataArray_t<real, 1, 1> Prandtl_SpecificHeatPressure = {{ 2.5 }} ;
  }} ;

TurbulenceClosure_t TurbulenceClosure =
  {{
  TurbulenceClosureType_t TurbulenceClosureType = EddyViscosity ;
```

```
    DataArray<real, 1, 1> PrandtlTurbulent = {{ 0.90 }} ;
    }} ;

  TurbulenceModel_t<3> TurbulenceModel =
    {{
    TurbulenceModelType_t TurbulenceModelType =
      OneEquation_SpalartAllmaras ;

    int[6] DiffusionModel = [0,1,0,0,0,0] ;
    }} ;
  }} ;          ! end FlowEquationSet
}} ;           ! TwoZoneCase
```